HYDROLOGICAL FORECASTING

WATER SCIENCE AND TECHNOLOGY LIBRARY

HYDROLOGICAL FORECASTING

Design and Operation of Hydrological Forecasting Systems

by

JAROMIR NEMEC

Director, Hydrological and Water Resources Department,
World Meteorological Organization

D. REIDEL PUBLISHING COMPANY

A MEMBER OF THE KLUWER ACADEMIC PUBLISHERS GROUP

DORDRECHT / BOSTON / LANCASTER / TOKYO

Preface

In the past two decades several activities in the field of water resources management have been enhanced and intensified. This rise had at least two independent reasons. The first and main one was the constantly increasing water demand for agriculture and industry on one side and the concern about the deteriorating environment on the other. While this last concern was lately overshadowed by deterioration of national economies, the quantity of available water resources has certainly not increased with the growing scarcity of funds for its development and protection. Furthermore, the standard of living, which raised across the world, even in India and China, countries which concentrate more than a third of the world population, has made people and their governments more aware of natural disasters caused by weather. Since a large percentage of losses in human life and material damage from weather-related disasters are caused by water, either by its excess or scarcity, the concern about water has been increasingly associated with these disasters.

The second reason for intensified water resources management is man's spectacular technological advance in electronics, computers and use of satellites. The Koran says that two things cannot be predicted: the sex of the child in its mother's womb and the quantity of water that falls from the sky and flows in rivers. Technological progress has disproved both of these caveats. It was suddenly recognized that the management of water resources and mitigation of of disasters caused by water will be more efficient and economical with the knowledge of future availability or scarcity of water. This knowledge is acquired by hydrological forecasting, which was until then considered as a very poor relative of hydraulic structures - in particular dams, dikes and other investment intensive measures used for management of water resources. Results of hydrological forecasting having been improved by the new technology, many countries of the world started to establish hydrological forecasting systems (HFS). But the field is new, there is very scarce, if any, comprehensive guidance on the design and operation of such systems.

The author of this book was called upon to advise the design and operation of such systems in countries of the third world in the late sixties and with the exception of a few in-service texts available in USA and USSR, there was no recorded relevant experience. It was a process of learning in practice, both for the author of this book and the handful of experts who were in charge of the projects of the World Meteorological Organization assisting the developing countries of Latin America, Africa and, in particular Asia, most often financed by the United Nations Development Programme (UNDP). This book is a record of the experience acquired in these projects as well as in these of national services of countries, which were involved in the establishment of the hydrological forecasting systems on some of the most important rivers of the world. It is designed to help specialists in all countries, but in particular those working in the developing countries.

It was not the author's intention to replace or duplicate other material published on this relatively new subject by WMO or elsewhere. Nevertheless, in order to enhance the publication's usefulness, making it self-contained, excerpts from such material were included with a clear indication of the source. In particular some parts of the WMO Guide to Hydrological Practices (WMO, 1983) and of several WMO Operational Hydrology Reports have been included since they are considered very pertinent. This book may serve as a textbook for the student, a manual for the general practitioner and a reference for the specialist. Examples of HFS from projects executed by WMO, in which the author was involved, enhance the practical aspects of the book. In many cases experience gained was put for the first time on record for readily available reference, with the hope of multiplying in this way the benefit from the funds invested in the established HFS.

Last, but not least, the experience recorded is that of people from many countries of the world who, without distinction of their colour or mother tongue were united by the idea that they are helping through their science and technology to improve mankind's well being and mitigate its suffering.

The author wishes to acknowledge with gratitude the kind permission of the Secretary-General of WMO to make use of WMO material and to submit this text for publication. He further acknowledges the assistance and advice he received from many colleagues, both in the WMO Secretariat and in the Hydrological Services of different countries, in preparing this publication. To all of them he expresses his sincere thanks. Last, but certainly not least, he wishes to express his gratitude to the colleagues and in particular to Ms. Corazon Espejo who made the text suitable for reproduction.

1. Introduction

1.1 DEFINITION OF HYDROLOGICAL FORECASTING (HF)

Hydrological forecasting is the prior estimate of future states of hydrological phenomena in real time, not to be confused with hydrological design-data computation (prediction)*. While the actual use of the words "forecast" and "prediction" is a matter of semantics or convenience, the substance of the technical activities connected with each is basically different. Both these activities require historical data (time series) and may use the same or similar methods of hydrological analysis and modelling. However, hydrological forecasting comprises additional technical activities connected with other hydrological and non-hydrological subjects, such as network design, data processing, remote-sensing techniques, telecommunications, operational use of computers, etc. In view of this, the subject of hydrological forecasting should not be viewed as one particular hydrological technique, but as an economic activity using many technological developments, both hydrological and non-hydrological. Thus a prior knowledge of hydrological analysis and modelling, of network design techniques, of telecommunications and remote-sensing systems, of on-line and off-line computer use, etc., is necessary if the activity of hydrological forecasting is to be exercised successfully and efficiently.

The difference between "forecasting" and "prediction" is often also of significance for the professional hydrologist or other specialists engaged in one or the other activity. The hydrological analysis procedures (models) used for both may be the same, but the reaction of the layman and/or specialized community to the result of their work is basically different. Indeed, the design hydrologist (engineer) advances a hypothesis - to the best of his professional knowledge - but only a hypothesis (of a storm or flow occurring up-stream) and not often, if ever, does anybody expect that this hypothesis will occur exactly as assumed. If the subsequent analysis and its results, as pointed out by Sittner (1982), is loaded with uncertainties and inaccuracies, no one will hold him responsible for it, unless a blatant professional mistake or dishonesty is committed. For the forecaster, who starts with an actual occurrence (storm or flow), the result of his work will be

*The word "prediction" is used here to distinguish between forecasting in real-time and computation of data for structure design purposes. It is a purely semantic distinction, as "forecasting" and "prediction" have the same general dictionary definition.

In October 1985 an international conference was held in Villach (Austria) sponsored by the United Nations Environment Programme (UNEP), the World Meteorological Organization (WMO) and the International Council of Scientific Unions (ICSU). It considered the role of carbon dioxide and of other greenhouse gases in climate variations and associated impacts. The Conference reached a conclusion that if present trends continue,the content of CO_2 in the atmosphere will double and according to the models available, this would increase the global mean equilibrium surface temperature in the range of 1.5-4.5°C by the beginning of the next century. Regional scale changes in climate have not yet been modelled with confidence. From regional modelling results it appears that the possibility exists for the annual mean runoff to increase in high latitudes and summer dryness to become more frequent at middle latitudes in the Northern Hemisphere. The Conference pointed out that based on evidence of effects of past climate changes, there is little doubt that a future change in climate of the order of magnitude obtained from climate models for the doubling of the atmosphere CO_2 concentration could have profound effects on global ecosystems, agriculture, water resources and sea ice.

The above statement indicates that many decisions being made today on long-term projects, in particular major water-resource management activities in agriculture such as for irrigation and drought relief, cannot be based on the assumption that past climatic conditions, without modification are a reliable guide to the future. Several papers presented at the Conference, in particular this of Warrick et al (1985) were concerned more specifically with the effects of the predicted climatic change on agriculture. The paper stresses that this impact will be on one side caused by gradual change in average (regional) climate, on the other by a change in climatic risks. The two aspects present a different picture from a point of view of dangers or benefits. The first results is a slow deterioration (or enhancement) of the growing environment. The second causes changes in the frequencies of unusually disruptive (or beneficial) climatic events. This latter aspect does not challenge the importance of a slow regional climate change, but suggests that since this change will be slow, agriculture will have some time to adapt. On the other hand the "shift-in-risk" aspect stresses the need of increasing attention to be paid to existing methods of assessment of risks arising from inter-annual variability of climate, which may, due to climatic change, become much more pronounced in all regions of the world.

The potential for a dangerous succession of climatic extremes could thus escalate rapidly, while the climate change as measured by the central tendency itself were very slow. Warrick et al (1985) point out that in particular some representatives of developing countries claim quite rightly that the slow change in the mean "is quite secondary to the present problem of inter-annual variability", the shift-in-risk perspective in the agricultural impacts of climatic change could be interpreted as a potential severe exacerbation of existing problem with "yields (of crops and water)" - a problem which could be felt acutely, abruptly, and possibly in the not-so-distant future.

From the above discussion of state-of-the-art research on the impact of most probable impending climatic change on agriculture and water resources a few important points concerning hydrology and water resources management may be inferred. The increase of the variance and amplitude of both the extremes of availability of water has particular incidence on storage, the only way of redistribution of water in time and to a lesser extent on redistribution of water in space. The basic results of climate modelling for global averages indicates doubtlessly that the time redistribution of extremes and hence increased need for storage is to be expected to occur with high probability. The modelling of regional effects is yet not as advanced (or trustworthy) as it would be wished, but no doubt redistribution of water in space, assuming the spatial meso and macroscales of climate change, is a less acceptable proposition than redistribution of food supplies. In simple, although perhaps not entirely satisfactory words, it is easier and cheaper to transport grain than water on long distances, particularly across seas or oceans.

It is to be expected that the role of hydrological forecasting in real time, and in particular of long-term and medium-term forecasts for water management in general, as well as short-term forecasts for reservoir operation for irrigation purposes will be gaining on importance.

1.4 CLASSIFICATION OF FORECASTS

From the above general discussion, it follows that hydrological forecasts can be classified mainly by three, mutually interdependent characteristics:

(a) The forecast variable;

(b) The purpose of the forecast ;

(c) The lead time, also known as forecasting or forewarning period.

Thus forecasts can be classified as follows:

(a) *By forecast variable:*

- (i) Water levels (rivers, lakes and reservoirs);
- (ii) Water quantity (discharges, volumes) and quality;
- (iii) Ice formation and break-up.

(b) *By purpose of the forecast:*

(i) Flood forecasts (including flash floods, dam breaks and storm surges in estuaries and coastal areas resulting from combined river and sea flooding);

(ii) Water supply forecasts for different periods (daily, weekly, seasonal, annual), including forecasts for water power production and irrigation;

(iii) Navigation forecasts;

(iv) Agricultural forecasts (mainly river levels for seeding, harvest or other purposes);

(v) Low-flow forecasts for pollution control.

(c) *Forecasting period:*

Short-term hydrological forecast:	Forecast of the future value of an element of the régime of a water body for a period ending up to two days from the issue of the forecast.
Medium-term (extended) hydrological forecast:	Forecast of the future value value of an element of the régime of a water body for a period ending between two and ten days from the issue of the forecast.
Long-term hydrological forecast:	Forecast of the future value of an element of the régime of a water body for a period extending beyond ten days from the issue of the forecast.
Seasonal hydrological forecast:	Forecast of the future value of an element of the régime of a water body for a season (usually covering a period of several months or more).
Hydrological warning:	Emergency information on an expected hydrological phenomenon which is considered to be dangerous.

The above classification is all comprising, in many cases there is sufficient to clasify HF in four categories:

1. Flood forecasts

2. Medium and long-term forecasts of flow (including low flow forecasts)

3. Forecasts of ice formation and break up

4. Forecasts of water quality

It is obvious that one of the most important criteria of the classification is the lead time. Its definition is not universally accepted, but a large consensus can be assumed around the following one: The lead time is the time between the issuance of the forecast of a hydrological phenomenon and its occurrence. This lead time may be considered as the net lead time, a gross lead time would be the time between the occurrence of the cause(s) of a hydrological phenomenon and its occurrence (i.e. between the occurrence of a rain causing a flood and the flood peak occurrence). The use of the gross lead time is nevertheless not recommened, as there is rather large uncertainty about the times of occurrences of causal phenomena. The relativeness of time occurrence with space occurrence has to be also taken into account, as in flood forecasting in particular, a relatively short lead time for a short river reach may become a long lead time for points much further downstream. Fig. 1a illustrates the above concepts.

1.5 RELATIONSHIP OF METEOROLOGICAL AND HYDROLOGICAL FORECASTS AND RELEVANT SERVICES

Prior knowledge of the meteorological conditions increases the scope and efficacy of hydrological forecasting, lengthens the validity of such forecasts and increases their accuracy and reliability. Quantitative precipitation forecasts (QPF) and other meteorological forecasts (temperature, wind and snow conditions) constitute an important and essential input to the present and future procedures and methods of hydrological forecasting. Considering the present trend in making use of rainfall/runoff models to simulate the catchment response to precipitation inputs, any improvements in timeliness of hydrological forecasting for any but the very large rivers will hinge chiefly on progress made in rainfall forecasting. The input to these models, usually the rainfall quantity during a given time period, is normally obtained from the actual observations and for most rivers the flood forecast from the model based on this input is but a few hours ahead of the actual event. As a consequence, the alert of flood-plain residents and subsequent remedial measures for their protection are under a severe time constraint in their execution. The alternative of taking remedial

problem, associated with river flooding, is emphasized by the fact that nearly 90 per cent of lives lost through tropical cyclones result from drowning in ocean storm-surges and floods. WMO was called upon by the international community to mobilize scientists and resources to discover ways of mitigating the harmful effects of these storms and of removing or minimizing their destructive potential. WMO responded by setting up its Tropical Cyclone Project. The main objectives of this project include the strengthening of present capabilities of detecting, tracking and forecasting tropical cyclones, the improvement of tropical cyclone warning systems and the strengthening of flood-forecasting capabilities. It will be referred to in several parts of this book.

The interdependence between meteorological and hydrological forecasts has, in many countries, led either to entrusting hydrological forecasting to meteorological services (for example in USA, Australia, Pakistan, the Philippines) or to a close collaboration between meteorological services and hydrological services or agencies in charge of rivers (river basin organizations) (for example in UK, India). As long as the units responsible for meteorological forecasts over a river basin are either part of the same organization or co-located with those in charge of the hydrological forecasting in the river basin, the use of meteorological forecasts in HFS does not present special problems. Often, however, this is not the case and the problem arises both of institutional co-operation and of physical communication lines. The problem is further complicated by the fact that, while general forecasts and charts may be available from national or even regional meteorological centres (under the WMO World Weather Watch, such centres have been established in all parts of the world), HFS activities necessarily require interpretation and correction of the general meteorological forecasts and charts for local conditions.

The interest for it was particularly stressed by countries in Asia, which in a regional effort supported by funds from the UNDP have made a spectacular progress in HF technology. In order to make available to national agencies in charge of HF in Asia an easily transferable HFS, WMO selected several appropriate elements of a HFS and connected them in a "sequence". This sequence is presented in manuals and computer programs for microcomputers. It will be referred to in several parts of this book.

1.6 USE OF HYDROLOGICAL OPERATIONAL MULTIPURPOSE SUBPROGRAMME (HOMS) OF WMO IN HFS

It will be recalled that the Hydrological Operational Multipurpose Subprogramme (HOMS) is intended to promote the transfer of hydrological technology between Members of WMO for use in their water-resource projects. It does this by making the technology available to users in the form of components. These components are of various kinds, for instance, manual and computerized techniques for data collection, processing and analysis; commonly-used hydrological models; manuals describing field or office procedures; or instrument specifications.

The objectives of HOMS are:

- To improve the quantity and quality of hydrological data available for use in water-resource and other projects;

- To assist directly such field projects of countries;

- To aid in the application of appropriate technology and in related training, especially in developing countries.

The functioning of the system is described by the HOMS Reference Manual which includes summary descriptions of all the components at present available through HOMS and advice on selecting and using appropriate components in water-resource projects and in hydrological forecasting.

The subject matter is laid out according to the major activities of an operational hydrological service, and is presented under the following headings: Policy, planning and organization; network design; instruments and equipment; remote sensing; methods of observation; data transmission; data storage, retrieval and dissemination; primary data processing; secondary data processing; hydrological models for forecasting; analysis of data for planning, design and operation of water-resource systems; mathematical and statistical computations.

Larger sections are further subdivided to give a comprehensive classification of the subject matter covered. This classification scheme, together with a degree-of-complexity indicator for each component, forms the basis of a component numbering system. Complexity is graded from the simplest manual calculation methods to the most complex, such as large computer-based rainfall/runoff models or sophisticated automatic instrumentation.

At present there are in HOMS some 350 components supplied by 26 countries. Additional components are being submitted and, once these are approved, they are issued as supplements of the HOMS Reference Manual several times a year. A list of components relevant to HFS design and operation is in Annex 1.

For the past three years over 800 transfers of technology components between 76 countries have taken place within the framework of HOMS. Those interested in this technology system may contact their HOMS National Reference Centre, normally established by the national Hydrological (in some cases Meteorological) Service, or directly the HOMS Office of the World Meteorological Organization (P.O.B. No. 5, Geneva 20, Switzerland).

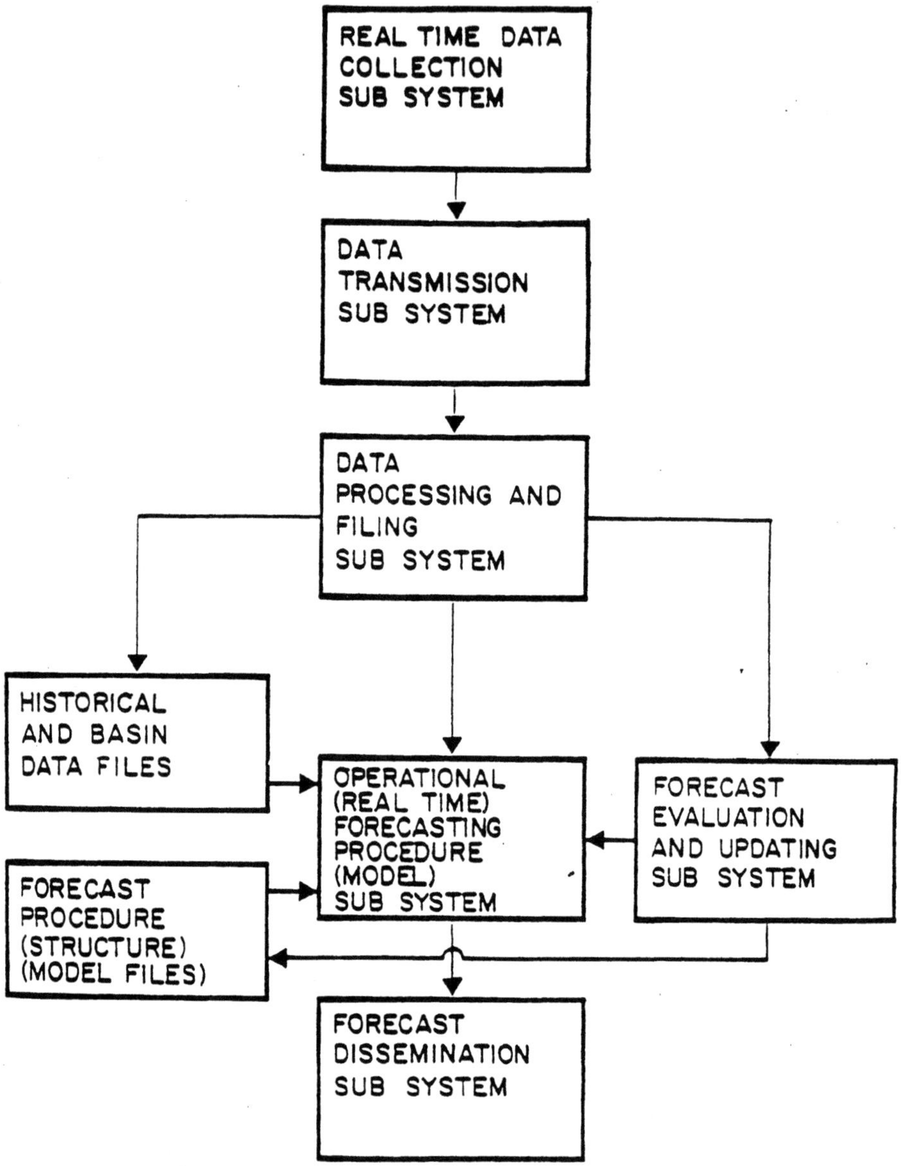

Figure 1b. Sub-systems of a hydrological Forecasting System (HFS)

2. Components of hydrological forecasting systems

2.1 GENERAL DESCRIPTION OF THE SYSTEM

A concrete HFS system depends on many conditions which, in addition to those of a technical nature and the natural environment (basin and river), include the social and administrative structures of the specific country. For this reason a general description of the design and operation of the components of a HFS can include only those sub-systems which are indispensable for the system in general. These sub-systems are:

- Historical and real time data collection
- Data transmission
- Data base management
- Forecasting procedure (modelling) in development and operational mode
- Forecast dissemination services
- Forecast evaluation and updating

The different sub-systems and their interdependence are illustrated by the flow-chart in Figure 1b. The selection of any particular sub-system is dependent not only on the conditions mentioned above but also on the other sub-systems, which in many cases may already exist. This is particularly the case with the data-collection sub-systems, because the historical data will have to be collected from an existing network, generally some time before the actual HFS is put into operation. Similarly, an institutional set-up (for example legal status of a government agency responsible for forecasting) normally exists particularly in countries subject to regularly occurring, hydrology-related hazards (floods and droughts), and the introduction of new sub-systems of HFS has to take into consideration the existing set-up. For the sake of simplicity and because it is impossible to foresee all possibilities of interaction between the sub-systems, these will be described as optimum solutions to be adapted to existing conditions. The different sub-systems and their action in the HFS system as time-space interdependent is illustrated in Figure 2, valid specifically for flood forecasting.

On the time scale of Figure 2 and on the parallel scale of basin areas of forecast, elementary basins, average times of concentration are indicated, calculated and averaged from hydrographs of many rivers basins of Europe, USA, Latin America and Asia, corresponding to the time from the beginning of the runoff-producing event to the culmination of a flood wave (maximum flow).

From Figure 2, it is evident that historical data collection, transmission and processing HF sub-systems will be most important in

operation of the HFS (the forecasts themselves). The latter redesign is a conceptual one. Improvements in the system may result from technological innovations, such as replacement of old hardware with new one which in its turn calls for a different and better software. The replacement of existing hardware does not necessarily result from its malfunction or total destruction, it may be caused by end of services provided by the manufacturer or his representative and/or lack of spare parts. The author has witnessed that such developments occur within a time period of 5-8 years, thus a planned period of operation of main hardware in a HFS should not be unrealistically long, the above time limits of 5-8 years can serve as an indicative life span of one cycle of a HFS. Without maintenance this life span may be sharply reduced, particularly if the maintenance is provided from abroad, a very frequent case in developing country. In an extreme case in Africa such life span was reduced to two years. On the other hand, the author has witnessed life spans of HFS cycles exceeding ten years.

In this connexion, two terms often used in HF system design, in the selection of the different sub-systems, need clarification and, at least for this text, definition. These are: "on-line" and "off-line" systems, and "lumped parameter" models and "distributed parameter" models.

An "on-line" HF system, in its strictest definition, is a system in which all sub-systems, from data collection until forecast dissemination, are automated and directly coupled (interfaced), and the operator either only supervises it or intervenes if any failure occurs. A recent survey made by WMO in Europe (see Annex 2a) indicated that such systems are not being used operationally. The participation of man's judgment, and/or his performance of some of the activities in the sub-systems is not only necessary but desirable, notwithstanding the possibility of eliminating these by the use of sophisticated technology. For practical reasons, therefore, an "on-line" HF system will be considered as one which works with automated real time data collection, transmission, processing and modelling, even if between the different sub-systems, some of the interface is provided by man's intervention. Conversely, an "off-line" system would be defined as such if all the interfaces between the sub-systems were provided manually by man's intervention. Indeed, as ascertained by the above survey, the most difficult problem is to inter-link the components so as to make the system fully operational. Experience shows that, even if each of the automated sub-systems performs satisfactorily by itself, when coupled, a long period of de-bugging is needed and frequent breakdown of the system occurs, mainly at the interfaces. A pilot (testing) operation of several years is normally needed to provide satisfactory results. As expressed by Bobinski (1979), there could be nothing more deceiving than an "armchair approach" such as "the data are being collected automatically, the personnel are trained, the computers are running and therefore there should be no difficulties". Indeed it is just at that point that the real problems start. Yet, if the investment in HF is to be justified, the system has to be reliable and operational at any time, under the most adverse natural conditions, for which it has to be built. A system which works perfectly when the weather is benign, the roads dry and the radio-communications uninterrupted by magnetic storms, does not necessarily perform in the same way during heavy storms, flooded roads and difficult or impossible access to gauging stations. It is under these latter conditions that reliability has to be ascertained. The conditions for reliability are, of course, a matter of statistical evaluation, as one single temporary failure, for example of radio connexion, does not necessarily jeopardize either the system as a whole or its general reliability.

The terms "lumped" and "distributed" parameter system may also be interpreted in a HFS in several ways. Basically a "lumped" system considers that all input (rain, evaporation) and characteristics of the basin and its behavioural parameters are distributed in a spatially homogeneous way across the basin, varying only in time. In a "distributed" system the inputs and parameters of the basin are a function of the spatial location (geometrical co-ordinates). It is obvious that the hydrology of the "real world" is always a distributed system. However, the errors and practical difficulties of the data collection sub-systems do justify the representation of the "real world" by "elementary" basins, in which the inputs and parameters may be considered as spatially homogeneous and thus as "lumped" systems. It should be noted that the separation line between "distributed" and "lumped" representation of a natural river basin is very relative and often arbitrary, as it depends mainly on the density of the observational networks or other observational facilities available (including maps). Thus a basin of 10,000 km^2 may be considered as a lumped or distributed system, depending on the data available on the spatial variation of rainfall, evaporation, vegetation cover, soil and geology, and thus either as one or several "elementary basins", the outputs of which are combined into one. This relativity of size of an "elementary" forecasting basin has also to be borne in mind when considering Figure 2.

2.2 HISTORICAL AND REAL-TIME DATA COLLECTION SUB-SYSTEMS

The data used in hydrological forecasts can be divided into two groups: the first includes those required for developing the forecasting method, and the second includes the information needed to operate the forecast system.

Within the concept of a system, the data base used in the HFS is divided into two groups of data:

(a) the data (information) on the (real time) state of the system; and

(b) the data (information) establishing the "knowledge" base of the system.

In this concept a HFS represents what is known in the artificial intelligence theory as an "expert system". If we consider Fig. 1. the real-time data collection and transmission sub-systems generate continuously new information, which is absorbed by the operational data processing sub-system, used by the operational forecasting procedure and subsequently by the forecast dissemination sub-system. The data is filled in the historical data base, which is part of the system's knowledge base. This base provides the possibility of evaluation and updating of the product, thus improvement of the system by modification and restructuring, thus providing the self-learning of the system. The re-computation of the parameters of the models, which represent a part of the knowledge base, is either results of a better knowledge of the real word system or of a change in its behaviour.

The first group, encompassing the conventional time series as well as historical, hydrological information necessary for testing and evaluating the trial forecast procedure, also includes those constant basin and river characteristics, for example sub-catchment areas, land use, soil, channel dimensions and slopes, needed for pre-computation of the processes occurring over the basin and in the channels. The second group includes the hydrological and meteorological data specified by the forecasting procedure to characterize the state of the catchment transmitted in real-time immediately before the issue of the forecast. It may also include a measurement of the forecast element itself which may be used to monitor the forecast performance or update the forecast.

The reliability of a forecast procedure is directly related to the amount of data available for the development of the procedure, and to consistency, i.e., no change in relative values of the records with time. Care must be taken to ensure that there is no bias between the data used to develop the forecast procedure and the data used for operational forecasting. For this reason, consistency in the records is as important as quantity of data.

The networks of stations from which both the above types of data are to be provided may not be the same. In general the development of the forecasting procedure (model) requires historical data covering a longer time and more space than those needed for actual issuance of the forecast. Although network design is a hydrological technique of its own, the general principle for a HFS network may be summarized in two simple rules:

(a) For the development of the system, use as much of the data available as possible, providing it is homogeneous, consistent and relatively reliable (the reliability being often difficult to ascertain, particularly for floods and low flows).

(b) For the operation of the system, establish real-time transmitting stations as few as possible, the criterion being satisfaction of accuracy needed for the purpose of the forecast.

Two examples of techniques based on the acceptable error concept can illustrate the above principle. The first has been used by the United States National Weather Service, Office of Hydrology, to ascertain rainfall-gauge network density for operation of a flood-forecasting system, using the standard relative (%) error E of average storm rainfall over a catchment relationship to basin area A, thunderstorm frequency N and mean annual runoff in the basin.

$$E = 7.7(A)^{0.2}/(N)^{0.48}$$

Figure 3 represents this relationship in graphs. Details of the procedure are in WMO publication No. 324 (1972).

The precipitation network is of course in the foreground of network design, since an average rainfall input is normally the most sensitive input, providing that the flood forecast is not based only on stream-gauging data (see Section 3.2). The above standard error is ascertained from a general equation (US Weather Bureau, 1947)

$$E = \frac{C_v}{\sqrt{n}}$$

where

$$C_v = \sqrt{\int_A (X-P)^2 \, dA} \Big/ A/P$$

X = point precipitation for a given duration
P = areal precipitation for a given duration
A = area within which n raingauges are distributed at random.

Several authors have examined this problem, using either an autocorrelation function (Kagan, 1972; Zawadzki, 1973) or the Monte Carlo simulation techniques (Hashimoto, 1974).

From their work and that of US NWS, Ishizaki (1979) presented a table of relative error as a function of the area A, duration of the storm and number of raingauges n:

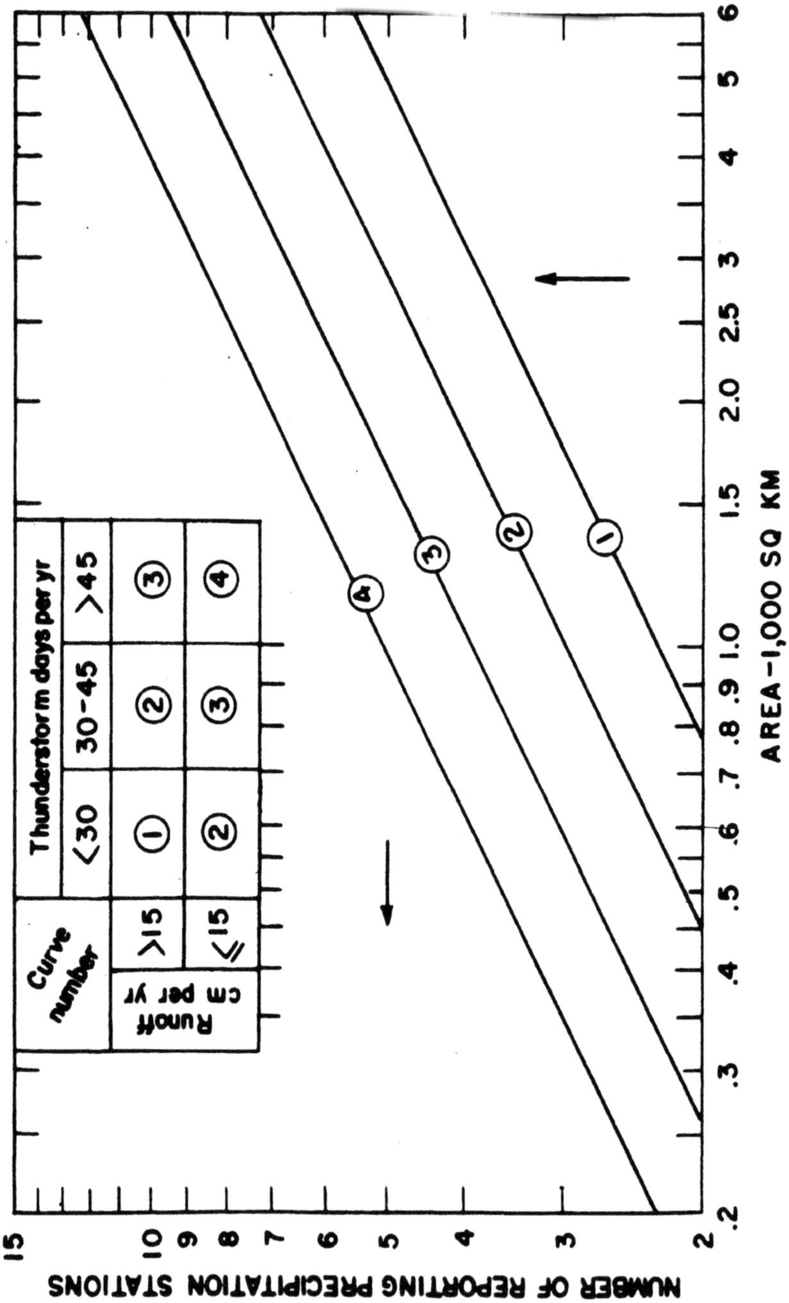

Figure 3. Network design for HFS.

Table I

Relative error of raingauge networks

Author: Duration of storm	Zawadzki 10^{-1} hour		U.S. NWS 1 hour		Hashimoto Total storm period	
n	10	100	10	100	10	100
A	Relative error					
100 km^2	0,39	0,08	0,21	0,07	0,08	0,02
500	0,55	0,12	0,31	0,09	0,11	0,04
1000	0,64	0,13	0,35	0,11	0,13	0,05
5000	0,92	0,25	0,49	0,16	0,20	0,06

Ishizaki concludes that a raingauge network is able to provide sufficient accuracy for flood forecasting providing the period of storm and/or forecast is not shorter than a few hours. Otherwise he considers that it is advantageous to use the radar technique discussed below.

A very logical approach to forecast raingauge network design was reported by Sugawara (1981). He uses a statistical evaluation of the spatial and temporal distribution of sampled showers in temperate and tropical climates, which he considers as a random, normally distributed process. He concludes that in temperate zones (Japan in particular) five raingauges are sufficient to provide a representative sample of the rainfall for forecasting purposes. In tropical zones this number is fifteen, independent of the size of the basin - assuming the forecasting technique uses a rainfall/runoff relationship in an "elementary" basin (of say 1000-3000 km^2).

As is demonstrated in Section 3.2, the accuracy of measured average rainfall, particularly for flood forecasting, is of crucial importance to the accuracy of the forecast. It is by far the single most important element of the HFS accuracy as a whole.

The network design of streamgauging stations for forecasting purposes is relatively simple for two reasons:

(a) The real-time reporting operational stream gauges have to be selected from existing stations, should a forecast procedure based only on correlation of upstream to downstream discharges be selected, or a rainfall-runoff model be calibrated by past observed data;

(b) New stations are selected predominantly at points for which forecasts are needed (the forecast points), so as to permit easy updating and evaluation of the forecast.

It should be noted with respect to the historical streamflow records that their accuracy is a function of their length in time and in its turn the stability of the parameters of a forecasting model is a function of the accuracy of the records. The network design based on an error function would, in this case, consist in the elimination from the historical record of those stations which may negatively influence the calibration of the model, providing that a choice of available record exists. An example of relationship between the length of record, its accuracy (or error) and different streamflow variables is indicated in Table II, valid for the State of Kansas in the USA (Langbein, 1972).

Table II

Variability and accuracy of statistical characteristics of streamflow

Streamflow characteristics	Standard error, in per cent for indicated length of record			
	Years			
	5	10	25	50
Mean annual flow	44	30	20	14
Standard deviation of annual flow	35	25	15	11
Mean monthly flow (average)	75	50	32	22
Standard deviation of monthly flow	35	25	15	11
50-year peak flow	87	58	36	24
50-year 7-day high flow	93	62	38	27
2-year 7-day low flow	78	53	32	23
20-year 7-day low flow	110	73	43	30

It is important to distinguish between the accuracy of a measurement or estimation (forecast) and precision which is a characteristic of the measuring instrument. There are specific requirements also for the precision of the data. As indicated by WMO (1983), recommended guidelines for the accuracy and frequency of data measurement are given in Table III, which specifies suggested goals for data measurement. The standards are recommended as guidelines to follow in planning equipment development and data acquisition programs for hydrological forecasting. Elements considered observable by automatic stations are also indicated. Rainfall, river level (or discharge, at a rated point) and water equivalent of snow depth are the most commonly required elements used by general-purpose forecasting centres.

Table III

Desirable precision of observation and frequency of data measurements for hydrological forecasting

Element	Precision[1]	Reporting interval	Measure by automatic land station
Precipitation -- Total amount and form[2]	2 mm below 40 mm 5% above 40 mm	6 hours[3]	Yes
River stage[4]	0.01 m	6 hours[5]	Yes
Lake level	0.01 m	Daily	Yes
Soil moisture	10% field capacity	Weekly	Yes
Frost depth	2 cm below 10 cm 20% above 10 cm	Daily	Yes
Water equivalent of snow on ground	2 mm below 20 mm 10% above 20 mm	Daily	Yes
Depth of snow cover	2 cm below 20 cm 10% above 20 cm	Daily	Yes
Density of snow cover	10%	Daily	---
Water temperatures[6] (rivers and lakes)	0.1°C in 0-4°C range otherwise ±1°C	Daily	Yes
Surface temperature snow	1°C	Daily	Yes
Temperature profiles (snow and lakes)	1°C	Daily	Yes
River and lake ice Thickness Type Character	0.02 m below 0.2 m 10% above 0.2 m	Daily	---
Water level (in wells)	0.02 m	Weekly	Yes
Net radiation	0.4 MJm^2/day below 8 MJm^{-2}/day 5% above 8 MJm^{-2}/day	Daily	Yes
Air temperature	0.1°C	6 hours	Yes
Wet bulb temperature	0.1°C	6 hours	Yes
Wind movement	10%	6 hours	Yes
Pan evaporation	0.5 mm	Daily	Yes

(1) With respect to actual observations the WMO Technical Regulations use the term "precision of observation or of reading" which is defined as the smallest unit of division on a scale of measurement from which a reading, either directly or by estimation, is possible.

(2) It may be necessary to distinguish solid and liquid forms of precipitation.

(3) Varies from one hour to one day, depending on river response. Event reporting for example, after 2 mm of rain required for flash flood forecasts.

(4) Depends on sensitivity of stage discharge relationship to stage change and can be 1 mm accuracy. If possible, an accuracy characterized by a relative standard deviation of 5 per cent should be arrived at.

(5) See note 3. Event reporting may be appropriate for flash flood forecasts.

(6) Hourly reporting with 0.3°C for ice forecasting.

The sensors installed in the stations collecting data in real time are in general the same as those in all other hydrological networks. Nevertheless the interface with the telecommunication equipment may, in some cases, require a particular type of sensor. Thus tipping-bucket raingauges are more common than float (siphon) types. While the nature of the river is decisive in the selection of the stage-recorder type, both bubble gauges and float-type gauges are being used. The new pressure transducers which are reaching the market appear quite promising as real-time stage-recorder sensors. Several manufacturers have their preferred sensors for instruments in packages of real time telephone or radio data-transmission sub-systems.

Remote sensing

The need for a detailed and rapid spatial coverage in real-time data collection gives particular importance to remote sensing of data. At this point it is important to emphasize the difference between remote sensing of data and their telemetering. In the former the sensor is always removed from the physical element to be sensed (rain, water level, snow cover, etc.). In the latter data from a sensor installed in situ are transmitted over a distance, sometimes even without recording the data at the location of the sensor (a solution which, incidentally, as a rule, is not recommended). According to WMO (1983), remote sensing from ground-based radar, satellite, and aeroplane, offers many distinct advantages for certain classes of forecasts, primarily because of the possibility of directly observing areally extensive variables which are otherwise only amenable to point sampling, and providing observations over inaccessible terrain or over the sea. It can be used to provide direct inputs to forecasting procedures in the following areas:

(a) Areal rainfall, both qualitative and quantitative indications;

(b) Areal extent of flood plain inundation;

(c) Cloud image indicating tropospheric wind as input to meteorological and hence hydrological forecast models;

(d) Tropical cyclone (or hurricane) movement;

(e) Area of snow cover;

(f) Water quality, in particular turbidity.

The following paragraphs describe the techniques, giving brief outlines of the properties of different equipment for various purposes.

Radar

The main uses of radar in hydrological forecasting are in assessing areal rainfall and in forecasting heavy rainfall and hence the resulting flood discharge, particularly in small catchments. For estimation of precipitation intensity and distribution, radars with peak power output of the order of 500 KW are used. The range in which hydrological computations can be attempted with such radars is limited to about 150 km in ordinary circumstances, but only 100 km at times of heavy rainfall. An additional point relevant to its use in forecasting is that although rainfalls of high intensity are measured most accurately by S-band radar, this band is more susceptible to the effects of screening and echo of the beam by surrounding high ground, and also to the effect of beam intersection with the melting layer (bright band).

Improved accuracy of quantitative estimates of precipitation depth can be obtained if electronic calibration of the incoming radar signals is carried out, based upon the regional relationships between radar and rainfall intensity for different seasons or storm types. This calibration is then applied to the radar image in real time with the help of computers using one or more interrogable raingauges.

Some hydrological forecasts require information on antecedent precipitation. This can be achieved with radar if facilities exist to store past values. This information, when combined with the drainage characteristics of the region, will help in locating the existence of any flood-potential area.

Recent developments in radar technology include the following:

(a) Computer graphics linked to radar to display storm activities on a radial grid (typically 0.3 to 1 km by 2° of arc) or cartesian grid (typically 2 to 5 km a side);

(b) Processing of digital data to provide subcatchment areal rainfall integration;

(c) Linking together of data from a number of radars by a telecommunication link to a central computer to increase the scanned area;

(d) Direct input of radar data into the forecasting model without further manual intervention.

(e) Use of doppler radar to forecast mesocyclones that precede thunderstorms and tornadoes. This requires a measure both of the echo reflection and the radial precipitation particle velocity.

Despite technical difficulties referred to, radar methods of estimating rainfall can have distinct advantages over other methods, particularly for flash flooding and for determining areal as against point rainfall. The greatest merit lies in being able to obtain a large amount of data on a real-time operational basis. However, to be able to derive these advantages, proper technical maintenance of the radar and associated ancillary equipment is essential.

Weather radar frequencies.

The frequency bands used in radar for precipitation observation are given below:

Weather radar frequency bands

Band	Frequency (MHz)	Wavelength (m)
S	1,550 - 5,200	0.193 - 0.0577
C	3,900 - 6,200	0.0769 - 0.0484
X	5,200 - 10,900	0.0577 - 0.0275

Precipitation attenuates the radar beam and this effect is greatest for short wavelength radar. On the other hand, long wavelength radar does not detect light rain and snow as readily as shorter wavelength equipment. The selection of a suitable wavelength depends on climatic conditions and the purposes to be served. All three of the radar bands given above are in use for observation of precipitation, the S-Band being recommended for tropical climates.

The radar-rainfall equation.

The radar equation is sometimes referred to as the FSMR equation (Free Space Maximum Range). This equation defines the maximum range to be anticipated from a particular radar system in terms of the design parameters incorporated with it and the effective target echo area. For precipitation targets where rainfall is considered to have filled the radar beam, the equation has this form:

$$P_r = \frac{P_t \pi A_r l [K]^2 Z}{8R^2\lambda^4}$$

where

P_r = average power in watts received from a series of reflected pulses;

P_t = peak power transmitted in watts;

A_r = effective area of antenna in m^2;

l = pulse length in metres;

R = range in metres;

λ = wavelength in metres;

$[K]^2$ = refractive index term of rain (.9313 for 10-cm radar equipment assuming a temperature of 10°C);

Z = reflectivity expressed as d^6 per m^3 where

d = drop diameter in mm.

The rainfall rate in $mm.h^{-1}$ is related to the median drop diameter as follows:

$$\sum d^6 = aP_ib$$

where P_i is the rainfall intensity in $mm.h^{-1}$ and a and b are constants. Many determinations have been made of the drop-size distribution measured at the ground and the conversion by means of the fall speeds of different sized drops to a particular rainfall rate. The most common equation in use is:

$$Z = 200P_i^{1.6}$$

Factors affecting measurements.

(a) *Precipitation type*

Because the return signal is influenced by drop size and is proportional to the sixth power of the diameter of the hydrometeor, it may be seen that a considerably stronger signal is received from the larger forms of precipitation. Hail, for instance, often gives an indication of intense precipitation, but snow yields relatively low reflectivity.

(b) *Beam width*

At 160 km the radar beam may be several kilometres wide, depending on the beam width employed. Normally, there will be marked variations in the radar reflectivity within this large sampling volume. Thus, an average value over a large volume is obtained, rather than a point value. The radar equation is based on the beam being filled with meteorological targets. Therefore, one would not expect the values of rainfall rate obtained with a radar to be highly correlated with point raingauge measurements.

(c) *Refraction of beam*

Radar waves are propagated through space with a refractive effect which gives the waves a curved path; the approximate mean radius is four-thirds the mean radius of the earth. As a result of vertical moisture discontinuities, additional refractive bending of the radar beam can occur. This produces what is often called ducting or trapping of the radar beam and either causes the radar beam to recurve earthward or to be curved upward overshooting precipitation 80-120 km away.

(d) *Atmospheric attenuation*

Microwaves are attenuated by atmospheric gases, clouds, and precipitation. The attenuation experienced by radio waves is a result of two effects: absorption and scattering. In general, gases act only as absorbers, but cloud and raindrops both scatter and absorb. For radar sets operating at the longer wave-lengths attenuation this is not a problem and can usually be neglected. The generally accepted form of expressing attenuation is in decibels.

The decibel (dB) is used as a measure of relative power and is expressed:

$$\mathrm{dB} = 10 \log 10 \frac{P_t}{P_r}$$

where P_t and P_r is the power transmitted and power received. respectively

(e) *Range attenuation*

The received signal power is inversely proportional to the square of the range to the target. Therefore, it may be seen that this is another form of attenuation which occurs as the beam propagates into space with the signal being attenuated as the beam width increases with range. The radar pulse energy in the beam is dissipated much like the light waves in a flashlight beam. There is electrical compensation for range attenuation on the display scope in many modern radars.

Methods and procedures of processing

(i) *Photographic methods*

For operational purposes an automatic camera with rapid development facilities is used to photograph the repeater scope to provide multiple exposure transparency photographs. These multiple exposures taken every 10 minutes are used to detect areas of persistent and/or intense rainfall during the last hour or two. The brightest areas on the film represent either the most intense precipitation or that of greatest persistence, where the echo reinforces the brilliance with successive exposures. After rapid process development of the transparency, it is mounted in a slide holder and the echo images are projected onto a map where they can be outlined and compared with observed rainfall reports.

(ii) *Manual techniques*

(a) Contour tracing

The simplest analysis involves periodic tracing of echo positions on transparent acetate radarscope overlays which have geographic boundaries outlined on them. Qualitative estimates can be made of the instantaneous storm intensity on the basis of echo appearance, weak, moderate or strong. Successive tracings of the scope display at 15-minutes intervals can be superimposed to determine the areas affected, the direction of storm motion, echo persistence, and the area-intensity relationship.

(b) Stepped-gain contour tracing

If the radar is equipped with a stepped-gain receiver control (attenuators) an intensity parameter can be introduced onto the radarscope overlay. Every 15 minutes for an hour the echoes at each of the selected dB attenuator levels are contoured. The use of different colours will readily show the most intense cores of rainfall as well as the areal extent of the storm.

(iii) *Automatic techniques*

An automatic electronic radar digitizer (DVIP - Digital Video Integration and Processor), capable of producing data recorded on magnetic tape in binary bits for immediate on-site computer analysis, transmission via data link to a remote computer, or for retention and later analysis. The results of this type of sampling are similar to those by manual methods, except that the number of discrete samples is larger, by at least one order of magnitude, than the finest grid overlay. The time required to sample and record the entire radar sweep is about 4 minutes.

The great "hydrological radar" controversy.

It should be pointed out immediately that radar provides information which is not obtainable from the classical raingauge network, namely a Quantitative Precipitation Forecast (QPF) and a detailed areal distribution of rainfall. Both these unique capacities of radar make it suitable for rainfall measurement for hydrological and water-management purposes. Here again, however, it is the user who has to state the accuracy necessary for radar to be useful and subsequently to what extent the benefit warrants the cost.

First, the radar has to provide at least as accurate and/or reliable information as the conventional network of raingauges. Secondly, the radar has to provide information for certain users readily usable on line, at least as easily as radio-telemetering raingauges. Thirdly, the information collection should be at a comparable cost to a network of raingauges.

In addition to the above-mentioned technical and economic considerations, which are valid in all countries, there are other factors which play a predominant role in the evaluation of the use of radar in developing countries. These factors are often underestimated or even overlooked when decisions are being taken on the installation of rainfall-measuring radar equipment for hydrological purposes in these countries.

To evaluate the requirement of accuracy it is necessary to analyze the errors inherent in the radar rainfall measurement. These are (Wilson, et al, 1979):

1. Errors in estimating the radar reflectivity factor;

2. Variations in the relationship between the rainfall intensity and reflected energy (the Z-R relationship);

3. The gauge/radar sampling differences (the $\overline{G/R}$ ratio).

The need to calibrate the radar by measurement of raingauges, which is at the base of the error indicated in 3. above, is caused by the fact that the Z-R relation, being accepted as constant (for example $Z = 200\ R^{1.6}$), is in reality not unique and varies during the measurement. There are other potentially serious sources of error not associated with the hardware, namely, radar beam blockage caused by obstacles, bending of the beam in an anomalous way, and attenuation by precipitation, which are most serious for shorter wave radars. Therefore, instruments used for precipitation measurement are different from the ordinary weather radars. Ten and 5.6 cm wavelengths are the most common ones.

The requirements indicated in 2. and 3. above, are best analyzed by a gauge-radar comparison. Table IV presents a summary of a comparison made in the USA with 14 storms, with gauges spread over an 8,000 km^2 area and at radar ranges between 45 to 100 km. The table shows that the ratio G/R varies from 0.41 (radar overestimate) to 2.41 (radar underestimate).

Several experiments were conducted to make adjustments by changing the Z-R relationship from storm to storm and even within storms, and calibrating the radar on a continuous basis with telemetering raingauges. Such adjustments evidently decrease the inherent error, according to Wilson et al (1979) to less than half (from 43-55 per cent to 17-35 per cent). Without the adjustments it should be reasonable to expect the radar measurements to be within a factor of two of the true rainfall (which can be only assumed), about 75 per cent of the time of measurement.

The 1979 review and summary of the state-of-the-art by Wilson et al is considered as a classical one, it is quoted by almost all authors writing on this subject since. There are a few others, in the United Kingdom (see Weather Radar and Real-time Hydrological Forecasting Project, 1977) and in the Federal Republic of Germany [Attmanspacher and Schultz (1980)]. The most recent ones are from the USA, USSR and Japan.

The paper from the USA (Hudlow et al, 1984) was designed to describe and justify the next generation of weather radars in the USA NEXRAD Project , with respect to quantitative precipitation measurement. It says: "Over 30 years of development and experimental activities still have not produced operational systems in the United States capable of providing consistently reliable quantitative precipitation estimates. There are several reasons for this, but the fundamental one has been the lack of an integrated systems approach to the problem. The operational feasibility has been established by the Japanese and British, who are ahead of the United States in many aspects of the use of radar precipitation estimates for water resources management."

The second study is from the USSR (Livanova N.A. et al, 1984). The study was conducted between 1978 and 1982 with a radar MRL-2, "Precipitation". The raw radar signal was used by a micro-coordinates with a grid of 10 x 10 km., with a subgrid of 3.3 x 3.3 km. for accumulation of the rainfall depth for 5 min, 1 and 6 h intervals. The working cycle of the radar was 5 min, the scanning interval was of 30 sec. The effective radius of the radar was 100 km. The calibration of the radar (derivation of Z-R relationship) was made on an experimental raingauge network of an area of 432 km^2 in the basin of the river Polometa at the Valdai Experimental station of the GGI (State Hydrological Institute). If the rainguages would have been distributed in regular intervals their optimal number should have been 30. Due to operational reasons, mainly those of access, this number was increased to 50. The error of the averaging of the rainfall depth by raingauges network was estimated to vary between 5-70%. This was used to estimate the error of the radar measurement of rainfall, which for this concrete case was 20-30%. A correlation of simultaneity of raingauge and radar data was also established. The coefficient of correlation r was 0.93 - 0.97.

Furthermore it was estimated the areal distribution of the rainfall fields measured by raingages and the radar. The raingauge data were interpolated into grid points with 0.55 km distance. The coefficient r between raingage and radar values was ascertained to

0.65-0.82, which was considered as unsatisfactory. It was assumed that the main factor of this discrepancy was the error in averaging of raindepth during the calibration period of the radar, bearing also in mind the mountainous character of the basin. In order to improve the calibration of the radar a secondary calibration was made using so called scaling coefficients. The secondary calibration, replacing the continuous calibration by telemetering raingauges, was made by three types of scaling coefficients.

$$M = \frac{R_{RA}}{R_{RG}}$$

and relative deviation

$$\underline{1} = \frac{R_{RA}-R_{RG}}{R_{RA}}$$

where

R_{RA}= average rainfall depth estimated by RA

R_{RA} and R_{RG} = rainfall depth in the elementary grid square of the RA, estimated by the RA and RG respectively.

Only values with a dry random were used to establish maps of scale coefficients.

A fourth estimation was made without any secondary calibration (no change in the Z-R relationship).

The comparison of these estimations has indicated that the average error of 44% without a secondary calibration was reduced to 27-28% with such a calibration.

This very thorough study ascertains errors very similar to the ones ascertained by Wilson et al.

The third study was made in Japan (Yoshino and Kozeki 1985). The results of it included in Table 5 are self-explanatory and corroborate the results obtained in the USA and USSR.

The maintenance of the radar and related computers in the conditions of the experiments described in the two latter above studies was no doubt very careful and no operational breakdowns occurred. The first study by Hudlow et al states, that to derive benefits from the radar for hydrological purposes the operational reliability should approach 100%. What is the situation in developing countries? WMO has assisted 34 countries to establish HFS on 17 rivers. In six of them a radar was installed. Three examples will be given:

Table IV

Summary of gauge-radar comparisons for 14 Oklahoma thunderstorms
(Based on 5-minute radar samples and $Z = 200R^{1.6}$)

Date	Number of gauges	Storm duration (h)	G (mm)	$\overline{G/R}$	Relative[1] dispersion about $\overline{G/R}$ %	Average[2] difference %	Average difference (Storm bias) %
28 Apr.1974	21	7	11	1.62	25	35	16
29 Apr.1974	20	12	7	0.82	21	30	17
29 Apr.1974	21	5	22	2.23	31	51	26
1 May 1974	21	9	24	2.41	20	57	13
20 May 1974	12	7	15	0.68	46	75	37
21 May 1974	10	3	5	0.59	36	87	27
23 May 1974	5	2	3	0.64	10	60	8
25 May 1974	22	6	25	0.91	27	30	27
25 May 1974	12	5	23	0.88	32	31	19
30 May 1974	22	7	25	1.09	34	32	33
3 Jun.1974	15	5	7	0.41	29	160	23
6 Jun.1974	13	2	16	0.68	43	79	36
8 Jun.1974	10	5	31	0.49	39	141	42
7 Apr.1975	19	4	14	1.13	23	17	18
				1.04	30[4]	63	24

[1] Relative dispersion (coefficient of variation) about $\overline{G/R}$ = 100% or $[G/R]/\overline{G/R}$; where $[G/R]$ is the standard deviation of the gauge-radar ratios.

[2] Average difference = 100% $\sum_{i=1}^{N} = \left| \frac{G_i - R_i}{G_i} \right| \Big/ N.$

[3] Average difference (storm bias removed) = 100% $\sum_{i=1}^{N} = \left| \frac{G_i - (\overline{G/R}) R_i}{G_i} \right| N.$

[4] The relative dispersion among all available gauge-radar comparisons (14 storms) is 63%.

Table V

Comparison of Errors between Ground Rainfall and Radar Rainfall after Yoshino

Intensity	o.1 mm/h or more		10 mm/h or more		20 mm/h) or more	
Cause of rain	Number of data	Error	Number of data	Error	Number of data	Error
All rain	2,747	178* 0.9**	210	51 7.4	38	32 25.6
Frontal	619	224 2,3	28	61 33.1	3	94 333.2
Typhoon	2,394	143 1.2	152	43 6.4	30	23 12.8
Thunderstorm	556	222 2.3	28	86 28.5	5	52 87.8
Cyclonic	178	144 1.8	2	36 49.9	-	-

* Upper values: relative error = $\dfrac{\sum_{i=1}^{N} \left| \dfrac{R_g - R_r}{R_g x100} \right|}{N}$

**Lower values: Square mean error = $\sqrt{\dfrac{\sum_{i=1}^{N} (R_g - R_r)^2}{N}}$

Rg: ground rainfall, R_r: radar rainfall

Example 1. Country X

Radar was a "C" band with narrow beam 0.9°, DVIP computerized by NOVA 3/12, 48k, 10 minutes scan cycle. Software existed for hourly accumulation, digital display and averaging in 17 basins with 16 x 16 km averaging square grid with telemetering raingauges. Data were transmitted by SSB radio over 100 km to flood forecasting centre. Practical results obtained have an error factor of 200% at radar site. Hardware and software failures occurred about 75% of the time of the flood seasons (four altogether). SSB transmission were functioning 10% of time of operation of radar. Calibration by raingages never operationally used due to failure of power supply (generators). At present radar not operational for flood forecasting. Price of radar was US$750 K (with all peripherals). Price of telemetering raingauges network of 45 stations was US$300 K.

Example 2. Country Y

The radar was a "C" band with narrow beam 1.2°. DVIP was computerized with DEC PDP 11/44. Accumulation was made by software for 15, 30, 60 min and 24 h. Scan every 5 min, map of 35 subcatchments. Software included data archiving/recording module, raingage reporting module and printout module. Raingauges reported via microwaves system and repeaters. Catchment was 150 km away from the radar location. Shipment scheduled for 1984. Building to house radar was ready in 1985. Technician of supplier arrived in June 1984. Lifting mechanism not available until end of July. Arcing of magnetron repeatedly made transmitter switch off. Defective modems did prevent to point catchment average rainfall. Ordered from 18,000 km away and delivered in two months. New magnetron installed two months later but arcing was not eliminated. When wave guide installed, arcing was eliminated. By then the flood season was over.

Computer (PDP 11/44) arrived before the radar. Could not be set into operation because of power supply problems. Two weeks later new part for power supply arrived. It was found out, when computer coupled with radar, that automatic calibration with telemetering raingauges does not work. Fault was in the software. The supplier has not yet developed it for this computer type.

In the last flood season (1985) the radar was not operational. Telemetering raingages not operational due to lack of spare PCBs. Price of radar and peripherals: US $1 M

Example 3, Country Z

For the past 15 years an "X" Band radar operates and provides qualitative information on precipitation for the flood forecasting center. The radar is mainly used to track typhoons.

Conclusions: ?

Compare requirements and conditions of US, USSR and Japan with those of country X, Y, Z.

To start with, in a developing country the very installation of the instrument can present difficulties of transport, and damage often delays the operational use of the radar for many months. Secondly, the maintenance of the radar is very difficult as, in the absence of a well-supplied stock of spare parts, the radar may be out of order for long periods of time, thus preventing any acquisition of data.

Finally, with respect to hardware, dust, heat and unqualified attendants are very difficult to eliminate from the radar room. The requirements for accuracy of data are also substantially different in most developing countries. For example, river-flow measurements, particularly during floods, serving either as historical data to calibrate the forecasting models or to update real-time forecasts, can be very poor, presenting errors of a factor of 2 to 3. Therefore, it is inconsistent, at least initially, to seek accurate areal rainfall measurement for the operational HFS.

Last but not least, the cost of a radar in terms of installation, maintenance and operation, as compared with telemetering raingauges, is rather overwhelming and if a telemetering system is necessary in any case, the purchase of an additional radar becomes of dubious benefit.

Satellite-based methods

Satellite data for hydrological forecasting is available from two types of satellites: geostationary and polar-orbiting. Geostationary satellites have an orbital period of exactly one day and, therefore, remain stationary at about 30,000 km above a point over the equator, whereas polar-orbiting satellites have an orbital period of between one and two hours. The usefulness of orbiting satellites for forecasting is limited by the frequency of the visit to the area of interest, the resolution and nature of sensing equipment and its sensitivity to any obscuring clouds. Because of these problems, the most established uses of satellite imagery are for long-term forecasts using snow-line detection, and for augmenting the synoptic interpretation given by meteorological forecasts.

Snow distribution and snow-line movement in mountainous basins can often be obtained in greater detail from satellite photographs than from available ground-station networks. Limitations on the use of satellite pictures are: an inadequate number of satellite observations, cloud interference, forest cover, and resolution of the satellite data. Quantitative analysis of snow depth is only possible when satellite observations are used in conjunction with ground observations. It has been found that present APT (Automatic Picture Transmission) satellite data can be useful in determining ice break-up for lakes and reservoirs over 500 km^2 in area.

In satellite pictures snow appears considerably brighter than the normal background tone of the snow-free surface. Therefore, snow-covered areas are easily identified in cloudless weather. Snow cover can be differentiated from cloud, both by the stability of its pattern as compared with that of cloud (for this purpose, it is necessary to have observations a day or more apart), and also because snow-covered areas and clouds appear to have a different texture (snow has a smooth appearance with sharp edges while clouds appear rough or lumpy with less defined edges). In mountainous regions the dendritic pattern of the river systems helps to identify snow cover.

The snow line encloses all areas which are identified, using the foregoing and other techniques, as being snow-covered. The snow line is defined as the boundary delineating the area with snow accumulation greater than about 2 cm. It is emphasized that the analysis should be based on all types of snow observations available. Comparison between satellite and conventional snow-course data shows that the snow line can be mapped from satellite photography with an accuracy of about 10 to 20 km. The accuracy depends on the scale of the picture.

Other meteorological and hydrological elements which may be observed or inferred by satellite-mounted sensors are: liquid water content of clouds and cloud pattern, areas and intensity of precipitation, flood-plain inundation or biomass loads and soil moisture.

Microwave sensors are used to measure the presence of moisture either in clouds, as falling rain, or an inundated ground area. Passive systems (which measure naturally emitted wavelengths) can provide "order of magnitude" estimates of precipitable water.

The most advanced use of satellite based observing systems for hydrological forecasting is at present in the USA, where these systems are combined with ground-based observations either from conventional stations or, for precipitation, from a radar network. The above-mentioned next generation radar network NEXRAD should contribute to make this type of data available for operational purposes in HFS. Many studies are at present underway for this purpose. One of the most complete is commissioned by the US National Aeronautics and Space Administration (NASA) under a program for Agricultural and Resources Inventory Surveys through Aerospace Remote Sensing (AgRISTARS). A series of reports pertaining to the use of remote-sensed data in hydrological models applicable in HFS has been prepared by Peck et al. (1983-1984) in the Conservation and Pollution section of the AgRISTARS program.

Meteorological forecasts are assisted by geostationary satellite data which, by allowing comparison between images taken as close as 30 minutes apart, provide inputs to numerical models and visual information about weather systems and hurricanes. Radiation temperature can be included in heat balance calculations for snowmelt forecasting.

The visual information provided by satellite imagery can provide invaluable information, especially in areas not covered by detailed survey, for catchment subdivision such as required when developing forecast systems. LANDSAT imagery is useful here and also where high frequency of observation is not required, such as in detecting water-quality changes in lakes and reservoirs, using its multispectral sensor.

However, it should be stressed that, with the exception of the snow line and of the snow depth, the acquisition of all other satellite remote-sensed data is at present only in the experimental stage. Even if rapid progress can be expected, it is not probable that a HFS designed between now and 1990 can incorporate an operational use of satellite-sensed Quantitative Precipitation Forecast, except in the continental boundaries of the USA.

This view of the author is corroborated by a recent survey made by a European group of hydrologists and water resources experts, supported by European Space Agency (ESA) on the use of remote sensing in hydrology and water resources. The conclusions of the survey were summarized as follows:

(a) Many operational requirements can only be fulfilled at present using traditional data collection methods (aerial survey included).

(b) Very few existing satellite data analytical techniques are seen as fully or potentially operational. Much seems to depend on the enthusiasm of individuals.

(c) The spatial resolution and frequency of information provided by existing sensors is inadequate for many applications. High incidences of cloud cover often exacerbate the imaging frequency problem.

(d) Satellite remote sensing facilities are expensive to instal, and new specialized staff may be needed to operate them.

(e) Knowledge and understanding of current and possible future satellite remote sensing systems and methods is low, implying needs for better publicity and education.

(f) Optimism for the future is greatest where there is access to digital image processing equipment.

Aircraft sensing

Many of the instruments and techniques used on spacecrafts can be mounted in aircraft. Because of the lower altitude the resolution becomes very much finer. However, the logistic problems of making an aircraft ready on short notice, and constraints imposed by cloud cover and reduced areal coverage, limit the usefulness of aircraft-mounted sensors particularly for flood forecasting.

Side-looking airborne radar (SLAR) has been used for terrain and flood-wave mapping and for monitoring lake-ice build-up and migration. Radar equipment is heavy, so light aircraft are usually not suitable carriers. Aircraft gamma radiation surveys have been used to determine the water equivalent of a snow-pack and, potentially, for soil moisture measurement. A limitation of the gamma technique is the low flying altitude (150 m) which is necessary.

2.3 DATA TRANSMISSION SUB-SYSTEM

The communication system transmits data from the field monitoring station back to the forecast centre. It may be manual or automatic, and it may report back at regular intervals or may warn when some specified event has occurred. The matter of networks has been discussed in Section 2.2 and is the subject of the WMO Casebook on Hydrological Network Design (WMO, 1974) and of the papers in WMO Operational Hydrology Report No. 8.

Important items in determining the means of communication required for an operational hydrological forecasting service are the density and location of the network of observation stations. Most data-transmission systems are also constrained by cost, availability of equipment, personnel, warning time, and communication channels. The manual and man/machine mix methods presently used to collect and communicate data are becoming inadequate for modern hydrological forecasting models and increasingly exacting user demands. Automation is one way to achieve faster collection and to obtain more frequent observations. Modern data-acquisition systems using computers for automatic collection of data on a predetermined schedule, or on the occurrence of prescribed changes in observed data, provide a relief from most constraints of manual methods, and provide the ability for expansion and evolution to meet users' increasing requirements.

The elimination of man as an observer and transmitter has some inherent disadvantages. Man has a great mental ability to integrate disparate information and supplements the purely factual and numerical report provided by the sensor with other environmental factors. Hence, the data from sensors in an automatic system may not be identical to those provided by an observer, and the success of an automated system is therefore quite dependent upon the availability of suitable and accurate sensors.

A few years ago a dilemma faced designers of HFS, particularly in developing countries. Should a fully automatic station, including a transmitter and not attended by man, be used, or should an observer be used, so as to avoid the failure of an automated station without its being noticed. This choice implied two economic factors: firstly, a fully automated station was more expensive than an ordinary radio (SSB single-side band short-wave transceiver); secondly, the elimination of the observer deprived a local inhabitant of an income, often the only one to be had. Two developments at least, in the view of this author, have contributed to the present solution to this problem. Firstly, automated stations of the third generation (reporting the change of the state of the hydrological elements and activated by this change or by a very simple timer), benefiting from the general price decline of electronics, became at least as low priced as a multipurpose radio transceiver (SSB), if not cheaper. Secondly, the reliability of such equipment increased to the degree that it surpassed the reliability of a local observer, at least of a part-time one who had minimal material motivation and often minimal education, if any. Furthermore, the possibility of easy repair of the automated stations (simple exchange of PCB-Printed Circuit Boards) made the operation of such stations dependent on a relatively less-qualified personnel. On the other hand, ordinary radio is often able to provide the only communication link and can also be used to disseminate the product of HFS - the final forecast and warnings.

Nevertheless, in existing and future HFS, particularly in developing countries, the following real-time data transmission sub-systems do and will exist:

- Manual;
- Semi-automated (mixture of man-operated and automatic stations);
- Fully automated.

From the point of view of purely telecommunication principles, the following systems can be used:

Direct wire

This system involves a hard-wire connexion or dedicated telephone line between the sensor and the central readout device. The readout device provides instant display of the sensor reading. An interrogator can also be used, programmed to interrogate automatically any one station, group of stations or the entire network at a variety of time intervals. An interrogator can be a teletypewriter or small computer terminal. The system may accommodate more than one parameter at each site and can relay the data to another location.

Low-voltage direct-wire systems are used to relay water-level recorder information from the gauging station to a convenient nearby location for ready access by the observer. These systems have a good history of reliability. The observer generally uses other means, such as the telephone, to relay the observations over longer distances to a collection centre.

These systems are not costly and are relatively easy to maintain. They are particularly useful in remote areas if the observer has a radio transceiver, whereby he can be in direct contact with a collection centre, relay point or forecast centre, depending on distance.

The availability of automatic digital recorders, memory devices and automatic relay equipment has permitted the automation of functions once performed by the observer. This approach has extended data availability to 24 hours a day. The system is more expensive than the one with an observer but less costly than providing the additional staff for the same coverage.

Direct-dial-telephone

This system utilizes commercially-available telephone circuits. Either a manual, automatic dial or computer-assisted dial system can initiate the outgoing call. Upon connexion the observational site responds and transmits current observational data and any data held in memory if the station is so equipped.

Data transmitted, depending on the sensor type, may be tones or other audible sounds. The sounds may be slow - slow enough to count and manually record - or they may be rapid, requiring special equipment to record and either display or print out the information.

With the availability of relatively inexpensive mini-computer systems, the use of commercially-available telephone circuits provides a rapid means of automating data collection directly at the observing site. This presupposes the availability of telephone circuits at or near the observation point and a certain degree of reliability of the telephone system, particularly in periods of storms and floods.

This system involves a minimum of investment in communication equipment. The major components are a mini-computer or specially-configured teletypewriter at the collection centre or forecast centre and a communications interface at the sensor sites.

Communication costs are the monthly tariff of the telephone company and the maintenance of the lines is their responsibility.

Line-of-sight radio

This system depends on very high frequency (VHF) line-of-sight transmission (usually less than 50 km) and does not therefore require relay stations. Greater distances can be achieved if the remote stations are elevated such as on a mountainside.

The network of remote stations can be designed to initiate a transmission whenever an event occurs. The receiving station can be equipped to process the data or relay them to a central processor.

Line-of-sight radio systems are less costly than radio-relay primarily because they are not dependent on relay towers and they offer benefits as well as limitations. One benefit is that the system is not compromised by the possible failure of a relay tower facility through which all or parts of the data flow. One limitation is that transmissions are usually restricted to line-of-sight distances, generally less than 50 km on level terrain. This approach, however, has particular value and application in relatively small basins where data may be collected from within the basin at a collection centre or the forecast centre.

More than one collection centre may be employed for larger basins and, in addition, this system can be extended for use by satellite. Data could be relayed from the collection centre to a satellite for further relay.

Radio relay

This system features radio equipment which includes the use of high frequencies (HF), very high frequencies (VHF), ultra-high frequencies (UHF), and super-high frequencies (SHF).

Super-high frequencies are generally used where multiplexed channels are required for data transmission such as from an Earth command station to a satellite. Ultra-high frequencies are used between Earth-orbiting satellites and data-collection platforms, as well as between relay stations.

In such systems the computer initiates an identifying signal which is transmitted from the data-control unit over a multichannel microwave link to a relay station, to repeater stations and finally to remote reporting stations.

The remote station responds to the proper identifying signal and transmits observational data to its assigned repeater station. Some systems are not interrogated but the transmitter of the remote station is activated either by a change in the state of the parameter observed (water level, precipitation) or by a timer (clock). Such systems have the advantage of not having a receiver at the remote station but feature only a transmitter. If the message is coded and transmitted very rapidly the consumption of power is minimal and the remote station can function for a very long period on a battery. This battery can either be changed from time to time (a period of six months is usual but not maximal) or "trickle" recharged by solar panels in locations where a sufficient insolation exists, which is the case of most developing countries with tropical and dry climates. The repeater station relays the data to the relay station and then to central control.

Satellite

As described above in the section on remote sensing, two types of satellite are employed for data collection: polar-orbiting and geostationary. A polar-orbiting satellite circles the Earth once about every 90 minutes and is therefore available for use as a relay station only while passing overhead within view of the Earth platform. A geostationary satellite is in a fixed location in space relative to the Earth above the Equator and, as it always views the same area of the Earth, is therefore available for continuous data relay. A smaller antenna and less power are required to transmit data to a polar-orbiting satellite because it is closer to Earth (about 1,000 km altitude). A larger antenna and more power are required to relay data to a geostationary satellite because of its higher altitude (about 30,000 km). A WMO co-ordinated system of geostationary satellites is represented in Figure 4. (WMO Publication No. 559, 1981)

Three basic types of data-collection-platform radio systems are most commonly used with both polar-orbiting and geostationary satellites: interrogated, timed and random access. With the interrogated type a call-up demand can be initiated. With the timed platform, reporting intervals are preselected and remain fixed at the site until changed. A third so-called "random access" type of platform transmits almost continuously and insures frequent observations. All data-collection platforms communicate with the satellite in the UHF band.

Data are received directly at the central readout facility and may be relayed over other communication facilities to various users. Details of each of the satellite systems are given in WMO Publication No. 411 (1975). The most important features, needed to be known by a HFS designer, before he approaches a telecommunication specialists are described below.

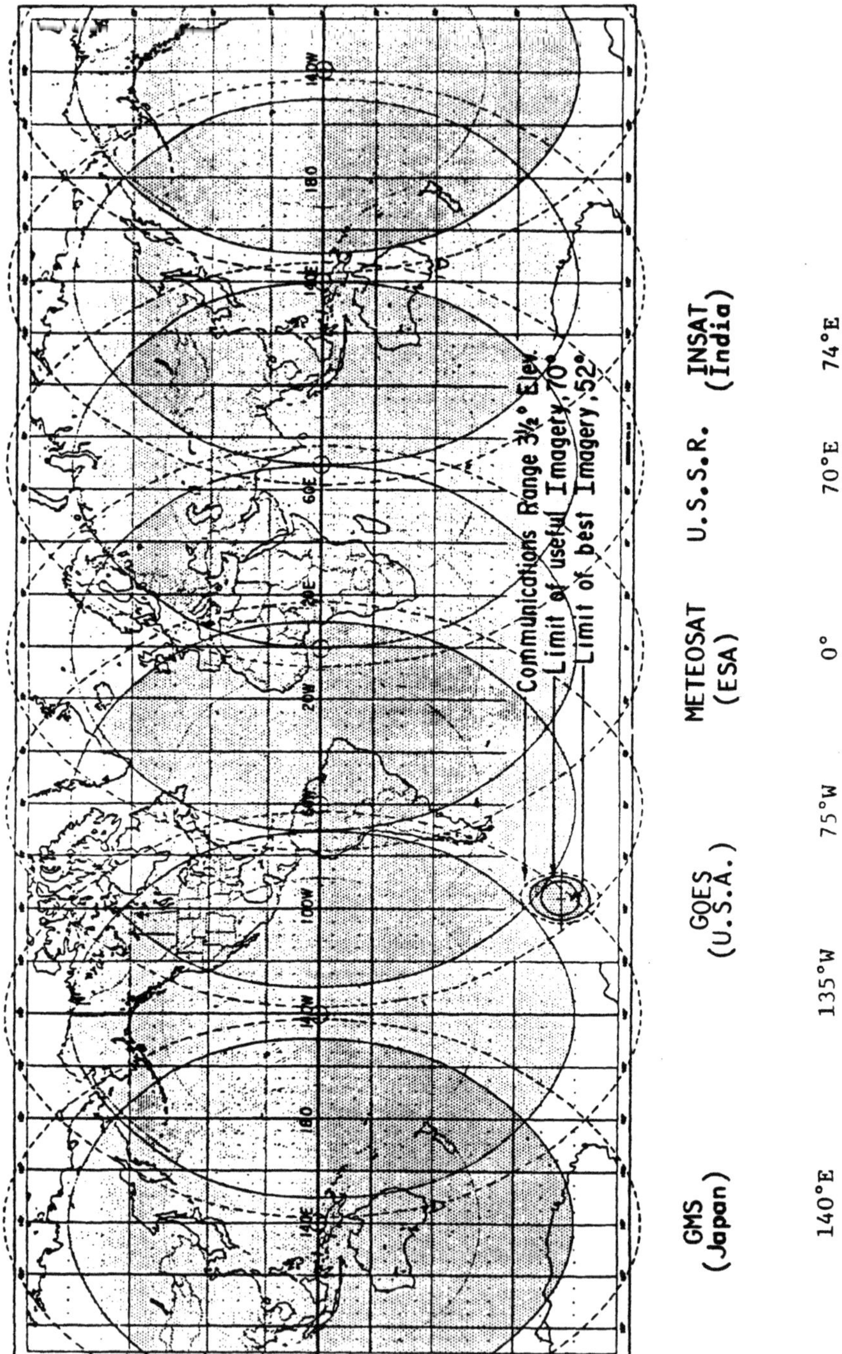

Figure 4. Geostationary meteorological satellites.

Geostationary Operational Environmental Satellite (GOES) of the USA

(i) *General*

With the launch in 1974 of the U.S.A. Geostationary Operational Environmental Satellite an operational environmental Data-Collection System (DCS) became available. The DCS includes a Command and Data-Acquisition Station (CDAS), the spacecraft which collects information from radio-equipped data-collection platforms and conforms to the applicable standards and regulations established by the International Telecommunications Union (ITU).

The use of the DCS is limited to the acquisition of environmental data. Environmental data are defined as observations and measurements of the physical, chemical or biological properties of the oceans, rivers, lakes, solid earth, and atmosphere (including space).

Users of the environmental DCS are responsible for the costs of the environmental sensors and data-collection platform radio equipment required to provide the communications link with the satellite, and any unique equipment/communications required to receive the data at the user's facility.

Design characteristics of the DCS on the spacecraft require that users conform to specific technical standards. An agreement is required that includes statements as to: (a) the period of time the agreement is valid and procedures for cancelling it; (b) conformity with ITU agreements and regulations; (c) required equipment standards; (d) standards of operation; (e) priorities for use; (f) reporting times and frequencies; (g) data formats; (h) data-delivery systems and schedules, and (i) user-borne costs.

Two of this series of geostationary satellites are currently in operation, one at 75°W to serve the Atlantic and the other at 135°W to serve the eastern Pacific Ocean. A third satellite is located near 105°W as an in-orbit spare.

(ii) *The GOES data-collection system*

The control facility for the GOES satellites is at the US National Oceanic and Atmospheric Administration (NOAA) Command and Data Acquisition Station at Wallops Island, Viriginiia, operated by its National Environmental Satellite Service (NESS). This facility has two 60-ft (18.3m) parabolic antennae, one for each of the two GOES satellites. A third antenna provides an emergency back-up reception capability in the event of a failure on one of the 60-ft antenna systems. The facility also has the necessary command, telemetry, data-processing, and communications equipment to interface with the spacecraft.

(iii) *Satellite Operations Control Centre*

In addition to performing telemetry control functions, the CDAS serves as the ground link between the spacecraft and the national US Satellite Operations Control Centre (SOCC) in Suitland, Maryland. The SOCC is responsible for the command programmes, which are transmitted by landline to the CDAS from which the personnel relay appropriate control instructions to the satellite. Imagery from the GOES data as well as telemetry on spacecraft status are received at the CDAS and relayed to SOCC where controllers analyse the spacecraft status and prepare the data for immediate relay to users or archive it for later call-up. Figure 5 (US Department of Commerce, 1977) is the overall outline of the CDAS, SOCC, and computer system for the data-collection system. System design will accommodate more than 10,000 DCPs within each six-hour period. The initial observing platforms consist of river and rain gauges, tide gauges, buoys, seismometers and automatic weather stations. The DCP radio set (DCPRS) transmit sensor data to the spacecraft in either an interrogated or a self-timed mode of operation (Nelson, 1975).

(iv) *Data-collection platforms*

The interrogated DCPRS (DCP Radio Set) is an externally actuated command-receiver system. The radio set is actuated by a command initiated by the CDAS and received through the GOES spacecraft. The DCPRS demodulates and decodes the received command. If the address agrees with the DCPRS stored address, the DCPRS is placed in the transmission mode in a manner identical to that of the self-timed DCPRS. Interrogation will be on a scheduled or as-needed basis. The programming of the synthesizer allows the radio set transmitter to be operated on any one of 100 selectable channels, plus one emergency channel. A lower-cost, single-channel DCPRS has been developed that uses a crystal oscillator instead of a synthesizer.

The self-timed DCPRS operates on an internally actuated, self-contained pre-programmed timer system. The timer is a clock that synoptically turns on the transmitter at one- to twelve-hour reporting intervals in one-hour increments. The operating frequency is selected at the DCPRS from any one of 50 channels.

Specifications for self-timed and interrogated typed DCPs are shown in Table VI.

A third type of DCP radio set is coming into wide use known as random reporting. It transmits almost continuously, thereby ensuring observational data frequently enough to meet most user needs. Random reporting was initially used with polar-orbiting satellites but is now being employed with geostationary satellites. The random-reporting DCP does not need the accurate time clock (oscillator) required with the timed DCP but does have to be

Table VI

DCP radio set specification (fixed location-type)

Parameter	Self-timed	Interrogated
Frequency		
Receive	N/A	468.825 MHz
Transmit	401.7 MHz to 401.85 MHz	401.85 to 402.0 MHz
No. of channels		
Receive	N/A	1
Transmit	50	100
Modulation technique	± 60° PSK Manchester	± 60° PSK Manchester
Transmitted power	5 W	5 W
Stand-by power dissipation	- 200 mW	- 500 mW
Receiver sensitivity	N/A	- 130 dBm
Antenna	Helix, right-hand circulated polarization	13 dB, 60° BW at 3 dB points
Supply voltage	12.5 V	12.5 V
Temperature range	-20°C to 50°C	-20°C to 50°C
Timing	1- to 12-h reporting intervals in 1-h increments; may be set to any 30-s period in the 12-h interval.	N/A

operated on a dedicated channel on the geostationary satellite to avoid interference with fixed-timed reporting DCPs.

Through microprocessor electronics the timed DCP radio set can be programmed automatically to increase reporting frequency when an event occurs. For example, a DCP radio set is interfaced with a water-level recorder and scheduled to report once every 24 hours. The river rises to a predetermined level, and upon exceeding this threshold the microprocessor instructs the radio to increase its reporting to a preselected, more frequent, interval such as every three hours. When the river falls below the threshold the microprocessor again instructs the radio to return to its once-a-day reporting interval. A self-timed DCP operating in this mode would be required to operate on a dedicated channel similar to the random reporting DCP to avoid interference with fixed-time reporting DCPs.

All these DCP radio sets can be operated on batteries which are "trickle-charged" by solar cells.

Frequency allocation

The GOES data-collection sub-system operates in the following frequency bands:

(a) 401 - 403 MHz - Earth-to-space link for data platforms;

(b) 468 - 469 MHz - space-to Earth link for platform interrogation;

(c) 1 690 - 1 700 MHz - space-to-Earth link for transmission of platform data to the CDAS.

Frequencies allocated by ITU for world use are shown in Table VII.

Geostationary Meteorological Satellite (METEOSAT)

The METEOSAT project was conceived to meet the requirements of the European meteorological community and to constitute Europe's contribution to the Global Atmospheric Research Programme and in particular to the First GARP Global Experiment (FGGE) and the World Weather Watch programme of WMO. Among the missions of METEOSAT of special interest to national Hydrological Services is the data-collection capability of the satellite that was launched in 1985. METEOSAT is at geostationary altitude at 0° longitude over the Equator. In this location, data collection is possible for all of Europe and Africa eastward to about 70°E longitude (Figure 5).

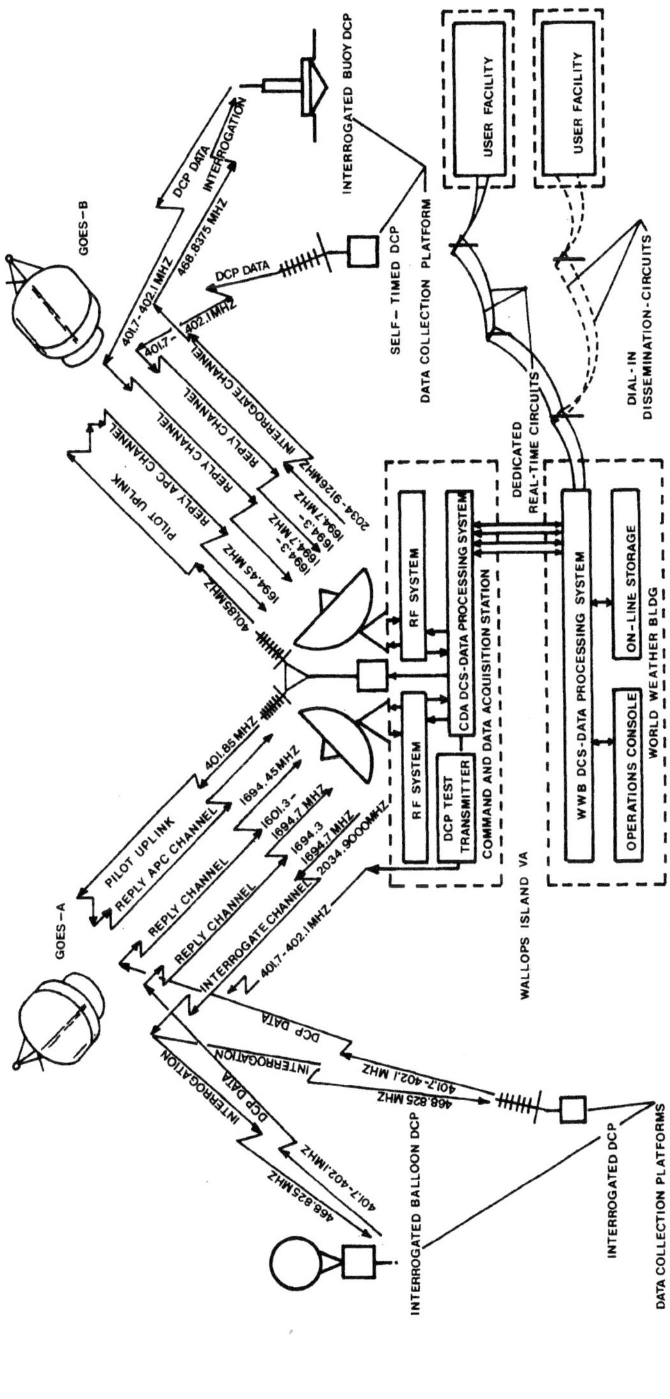

Figure 5. GOES data-collection system.

Table VII

Satellite communication frequencies as allocated by the International Telecommunications Union World Administrative Conference (Geneva, 1979)

Frequency	Use
137-138 MHz	Meteorological satellites (space-to-Earth) on a primary basis
400.15-401 MHz	Meteorological aids and meteorological satellites (space-to-Earth) on a parimary basis
401-403 MHz	Meteorological aids on a primary basis and for meteorological satellites (Earth-to-space) on a secondary basis
460-470 MHz	Meteorological satellites (space-to-Earth) on a secondary basis
1 670-1 700 HHz	Meteorological aids and meteorogical satellites (space-to-Earth) on a primary basis
1 700-1 710 MHz	Meteorological satelliltes (space-to-Earth) on a primary basis

The meteorological and hydrological programme office for METEOSAT is operated by the European Space Agency (ESA) in Toulouse, France.

Ground facilities for METEOSAT are at Darmstadt and Odenwald where the main telecommunication links with the satellite are located. These include the Data Acquisition, Tracking and Telecommand Station and the METEOSAT Operations Control Centre. The link to the GTS is via Offenbach.

Data-collection platforms for use with the METEOSAT system can be installed on land, at sea (buoys, ships) and aboard aircraft. There are two types of DCP: self-timed and interrogated.

International DCPs can operate in all zones covered by the various geostationary satellites whereas the regional DCPs will operate only with one satellite (Table VIII).

Table VIII

Meteosat data-collection system

Transmission of satellite interrogations to the DCPs		
	International (from 1981)	Regional
Frequency	468.875 MHz	468.925 MHz
Number of channels	1	1
Type of modulation	PCM-PSK Manchester ±60°	
Digital rate	100 bit/s	
Satelite E.I.R.P.	12 dBw	
Polarization	Right-hand circular	
Transmission of messages from the DCPs to the satellite		
	International	Regional
Frequency band	402 to 402.1 MHz	402.1 to 402.2 MHz
Number of channels	33	33
Channel width	3 kHz	3 kHz
Type of modulation	PCM-PSK Manchester ±60°	
Digital rate	100 bit/s	
E.I.R.P. of a platform	13 to 22 dBw	
Polarization	Right-hand circular	
Satellite receive G/T	-29 dB/K	

The ARGOS system

In addition to the GOES and METEOSAT geostationary data-collection systems experience, the polar-orbiting TIROS-N satellite with its ARGOS system is used for hydrological data transmission in HFS. One large HFS using operationally this satellite system is the Niger River Basin HFS in Africa.

ARGOS is a co-operative project between the Centre national d'études spatiales (CNES), Toulouse, France, and the U.S.A. The ARGOS programme started in 1978. The ARGOS on-board equipment (i.e., the receiving system) carried by the U.S. operational satellites designated NOAA-A to NOAA-G with a service of two in-orbit satellites planned until 1990.

The ARGOS on-board data-collection system is equipped to receive data transmitted by DCPs (fixed or moving) within the satellite's radio viewing range. Data are received on a random access basis. Under this arrangement the DCP in effect transmits its signals continuously (about every 100 to 200 seconds for fixed DCPs). As each message is acquired the time and data are recorded, the DCS measures carrier frequency and demodulates the platform identification number and sensor data.

These data are then formatted and stored by one of the on-board magnetic tape recorders. Each time the satellite passes over the telemetry station, the data recorded on tape are read out and transmitted to the ground receiving station

All DCPs transmit on the same frequency (401.650 Mhz). Duration of the messages is less than one second.

The spacecraft transmits data in real-time on 136.770 or 137.770 MHz.

Two systems are employed for relay of ARGOS data - one via meteorological telecommunication channels, the other via direct read out station.

The types of data-collection systems, and whether the satellite is in low or synchronous orbit, are shown in Figure 6 (Taillade-Carrière), including the operational satellite data-collection systems GOES, METEOSAT and TIROS-N.

Satellite system comparisons

Availability, reliability and cost are among the final set of determinants in the selection of a data-collection system. The user must be assured that the system will be available when needed. The user must also be assured that the system will be available when needed. The user must also be assured that the system is reliable

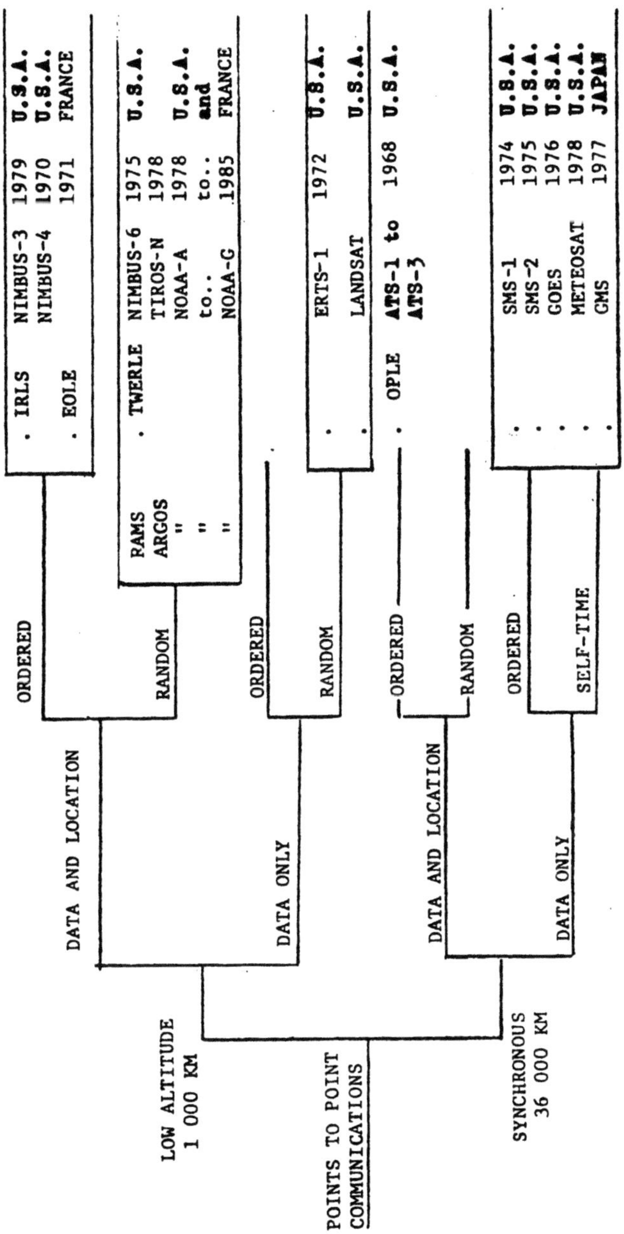

Figure 6. Data-collection satellites.

and not subject to frequent or long outage periods and that it will deliver the required data when needed. Finally, start-up costs, as well as operating and maintenance costs - including simplicity of repair - must be within acceptable limits.

Some comparisons have been reported indicating that geosynchronous satellite data-collection systems offer the most advantages for operational hydrology in terms of technical suitability and cost. One such comparison is the Canadian experience on the use of satellites for hydrometric telemetry. In Canada it is not always financially feasible to use conventional telemetry methods in remote areas. For example, the installation of power and telephone srvices at a gauging station site can cost over US $20,000 for each site.

The requirements of the Water Survey of Canada are being met by using the data-collection systems carried by the GOES and TIROS-N series of spacecraft. The Inland Waters Directorate of Canada has 100 time slots on GOES-east and 50 on GOES-west. No problems have been encountered in obtaining additional time slots nor have there been difficulties in renewing the agreement. A GOES receiving site is operated at the Prince Albert Satellite Station by the Canada Centre for Remote Sensing on behalf of the Directorate. The receiving site can be interrogated by means of standard 110- or 300-baud teletype terminals or by telex. The major drawback of the GOES system is that as the DCP antenna must be aimed at the spacecraft, there can be problems in obtaining line-of-sight in mountainous or northern regions. For example, in the Mackenzie delta, the antenna elevation angle is 10°. Another telemetry system must be selected for areas where use of GOES is not possible; the TIROS-N/ARGOS system is most appropriate. Bristol Aerospace, Ltd., in Winnipeg, is supplying GOES DCPs to the Atmospheric Environment Service. In addition, Bristol is developing a DCP that could be used with both the GOES and the ARGOS systems under contract with the Water Resources Branch. The DCP is designed to meet Water Survey requirements, and to survive in Canadian environmental conditions. The estimated cost, at US $5,000 for each unit, is about the same as the U.S. product.

The standard method of recieving GOES data is through NESS in Washington, D.C. Data are availalble in a raw format by means of 300- or 1,200-baud dial-in lines. A standard synchronous modem may also be obtained to distribute data automatically at rates of 2,400; 4,800; 7,200 and 9,600 baud. Most Water Survey of Canada offices have terminals that can be used with the 300-baud dial-in line; These terminals cost as little as US $1,200 and are suitable for persons operating small numbers of DCPs. Of the existing GOES users in Canada, the Province of Quebec has a terminal that has access to the 1,200 baud dial-in line. These terminals cost about US $5,000.

The data from TIROS-N can be received on much the same basis as that from GOES except that the distribution centre is CNES in Toulouse, France. It would be possible to install additional receiving equipment to handle TIROS-N data which could cost up to US $80,000.

Ground direct-readout receiving stations are being marketed for under US $50,000 that enable users to acquire data directly without needing to receive the information via other relay facilities.

Meteorburst

Meteor-scatter or, more commonly, meteor-burst communications rely on the ionization in meteorite trails to re-radiate or reflect radio waves in the low VHF frequency range for distances up to 2,000 km.

As a meteorite trail only exists for a few seconds, a burst (very short) transmission mode must be used. Also, the availability of ionized meteor trails varies with the time of day and the month of the year, requiring proper selection of operating frequencies.

The meteor-burst transmission system consists of at least two master polling stations plus the remote stations where the hydrological observing sites are located. The master station(s) can be linked to a central computer over dedicated telephone circuits if required. The system can operate unattended and an interrogation can be made for any one remote station or group of stations. Remote stations can also be powered by batteries charged by solar panels.

(i) *Master station*

The master station consists of a transmitter, receiver, duplexer, antenna, and a processor. A call-up message on a frequency in the VHF band is initiated to call remote stations. The master station can call any remote station or group of remote stations on command. In addition, the system can be instructed automatically to perform system-wide calling daily in the absence of other instructions and store data on-line as a back-up feature in the event that land-line outages prevent the forwarding of data. The master station can be computer-controlled and commanded over telephone circuits for unattended operation. The master station transmitter power output may be as low as 100 W but other systems indicate that power requirements of 1,000 W or higher may be needed, depending on whether one antenna for one quadrant coverage is used or four antennae for 360° coverage are used.

(ii) *Remote station*

The occurrence of a meteor trail in the correct geometric orientation, and of sufficient duration to allow the remote station to recognize its signal and transmit its data, is a random event. In the worst case, successful calling of a particular site can be expected to occur several times an hour, assuming the remote station is probed continuously. When a path has been established between the master station and the remote station, the remote station replies in the VHF band. Response time is about 100 ms. The remote station is controlled by a timer which activates the station at a predetermined time following a call-up message transmission.

The antenna system is a YAGI array, horizontally polarized and vertically mounted.

The number of times a day the remote station transmits a data message is determined by how often a valid call-up message is received and decoded by the remote station. The remote stations can be powered by batteries charged by solar panels. System design permits operation for up to a year without servicing depending on the type and variety of sensors employed.

(iii) *System applications*

The meteor-burst telemetry technique has many features which make it uniquely responsive to remote environmental requirements. The long-range paths which it creates will allow coverage of vast areas. Also, the use of meteor trails for signal reflection creates an unlimited number of potential paths for transmission. This minimizes, and in most cases eliminates, blockage problems due to terrain obstacles. Meteor burst also affords greater flexibility from the standpoint of expanding the number of remote stations. The addition of a site requires only the installation of the remote station equipment and the updating of computer tables in the master polling station. No hardware changes of any kind are required.

Another facet of the meteor-burst technique is that it allows complete control over which stations are called at which times. This control is effected by governing the scheduling and content of the master station probes.

The overall reliability of the system can be enhanced by the use of two master stations. During an outage of either master station, the system will collect data from all data sites, although significantly less efficiently. The second purpose of two master stations is to provide better-quality paths for communications. In cases where a data site location is disadvantageous relative to one master station, the other master station, or a combination of both master stations, will then yield the desired performance.

Data gathered show that meteor burst is suitable for hydrological data acquisition from remote locations. It may provide significant economic benefits, especially where conventional line-of-site radio links would require one or more repeaters. A meteor-burst system is at present being installed for hydrological and water-resource management purposes in Egypt on the River Nile.

2.4 DATA BASE MANAGEMENT

Data processing and data base packages

Part of this sub-system is normally carried out in the forecasting centre, but modern systems foresee some preprocessing at the collection site prior to data transmission. In fact, the design of HFS at present in preparation in several countries of the world foresees a considerable part of the processing taking place before transmission (due to rapid advances in microprocessors), thus achieving economy in data transmission. The entry of preprocessed data into computing units is easier although not necessarily more reliable as a man-supervised quality control is more difficult in this arrangement. Nevertheless man supervision is very important and should not be relinquished. Computer graphics are facilitating this supervision and control.

Operational data

The interface between the data collection, transmission and processing is the coding of the data, which depends on the processing needs. If several users require the data, in particular in international basins, it has also to satisfy some nationally or even internationally-agreed principles of coding, for transmission and processing.

WMO Hydrological Codes

To provide easily applicable hydrological codes, WMO, at the recommendation of its Commission for Hydrology, established two code forms, one for coding and transmission (and possible processing) of hydrological data, called HYDRA, and the second for transmission of forecasts, called HYFOR. They are included in the WMO Manual on Codes (Volume I, WMO Publication No. 306). The HYDRA code can be used to report:

- River or lake stage;
- Discharge;
- Precipitation and snow cover depth;
- Air and water temperature;
- Data on the state of ice on rivers, lakes or reservoirs.

The HYFOR code can be used to report:

- Stage forecasts and data-time of occurrence or data-times of the beginning and the end of the period for which forecasts are valid;

- Discharge forecasts with the same data-time specifications as for stage;

- Ice phenomena with the indication of the period for which the forecast is valid.

Station identification in WMO codes

The hydrological stations from which the data come in HYDRA code, or for which the forecast is valid in HYFOR code, are identified by two groups of five coded numbers as follows:

$$000\ AC_i \quad BBi_Hi_Hi_H$$

The first group is used only in transmission across international borders in international basins. The three 000 indicate empty space, filling only the group to five numbers, required by telecommunication regulations for code transmission, A = number of WMO Region (there are six such regions: 1 = Africa, 2 = Asia, 3 = South America, 4 = North and Central America, 5 = South West Pacific, 6 = Europe).

C_i = WMO Member country in that region. For national messages this group can be omitted.

BB = number of the river basin concerned according to an agreed list in the region.

$i_Hi_Hi_H$= number of the station in that basin.

The eight numbers thus identify the region, the country, the river basin and the station.

The number $i_Hi_Hi_H$ (which permits identification of 999 stations in each basin of one country) is assigned by each country. All details on the use of these codes are included in the above-mentioned WMO Publication No. 306 and also in WMO Technical Regulations pertaining to Hydrology.

Other codes used in HIS

In the processing of historical and real-time data for purposes of hydrological forecasting in more sophisticated on-line systems, the simple codes, such as those of WMO, will not satisfy the computer requirement and more versatile multi-purpose codes are devised. One such code, being introduced in the Columbia River forecasting system center at Portland, Oregon, in the USA (the originator of the SSARR model), designated as Standard Hydrological Exchange Format (SHEF) is designed to establish a data base including both historical and real-time data for the varied purposes of many agencies working in the basin (Columbia River Operational Hydromet Management System - CROHMS) (V.C. Bissel et al, 1982). Features of this code include:

1. It is readable by machine and man;

2. Describes a wide range of hydrological and meteorological data using two characters for each;

3. Makes compatible processing of raw observed data, processed products (data) and forecasts, all coming from a wide variety of sources;

4. The time identification of both time of observation and verification is acceptable in Greenwich and local time systems;

5. Identifies data from non-standard locations (for example bucket-surveys);

6. Allows for optional data descriptors for individual data, when included in specific files of the data base;

7. Permits different optional formating of data;

8. Permits use of comments within coded text to enhance readability;

9. Uses either SI or English units.

While it is doubtful that it will be possible to adopt one single code of such complexity for all HFS systems using different sources of data and processing facilities (computerized and manual), the tendency will be to standardize them as much as possible.

Transfer and processing of historical data

The processing of historical data is the subject of many standard hydrological procedures. Their general description is in the WMO Guide (1983) and in national publications; individual procedures are referenced in the HOMS Reference Manual (WMO, 1981) and are available through HOMS. To create files of such data for a HFS however, their transfer from other sources is necessary. The general rules for such transfer are given below (WMO, 1983). Details may be found in the WMO/FAO publication No. 634: "Guidelines for Computerized Data Processing in Operational Hydrology and Land and Water Management".

It is assumed that all such data have been preliminarily checked and subjected to error detection. Rules for such checks are again contained in standard national manuals or in WMO publications (WMO 1983). Their transfer may then be made on the following media:

(a) 80-column punched cards;
(b) Magnetic disk;
(c) Magnetic tape;
(d) Paper tape, 5 or 8 channel.
(e) Diskettes (floppy disks)
(f) Magnetic tape casettes (same as used by standard tape recorders)
(g) Computer to computer mail.

The volume of data to be transferred dictates which medium can be used in a practical sense.

Card readers are, however, no longer used and paper tape readers are also no longer usually available.

Because of their small bulk and speed of handling by the computer, diskettes and magnetic tapes are the preferred medium for transfer of all data sets.

Specifications for tapes

Differences in computer configuration make it necessary for the following conditions to be specified for transfer using magnetic tape:

(a) *Track details*

Magnetic tape may be written either in 7-track or 9-track form. Although the 9-track form is becoming the industry standard, some computers still use 7-track tapes.

(b) *Parity*

Odd or even parity of written information needs to be specified.

(c) *Density*

The density of information in bits per inch should be specified. The lowest density of packing provides the most reliable transfer. However, these cannot be read on modern tape drives.

(d) *System labelling*

Some computer systems demand that the initial record on the tape contain specific information. In most cases, however, unlabelled tapes can be handled.

(e) *Recording mode*

Because of differences in the word structure between computers, the data transfer should be as characters rather than in binary mode.

(f) *Code*

Various computers use different coded formats and the appropriate form should be specified. Even where the written code is not automatically recognized by the receiving computer, the information can be decoded by referencing to a dictionary record containing the alphanumeric characters used in the transfer. The bit pattern corresponding to each can then be recognized and the information be translated to the internal code of the receiving computer. The character packing should be specified.

(g) *End-of-file convention*

Most computers do not expect more than one end-of-file mark in any single data set.

(h) *Record length*

The length of the record in the transfer is recommended as 80 characters. If this is adopted, the file can be constructed of "card images", and any transfer undertaken can then be considered as being independent of the transfer medium.

(i) *Block length*

In the hardware handling of magnetic tapes, greater efficiency is achieved if groups of data records are blocked together. This length of the block in terms of an integral number of 80 character records must be specified.

Diskettes

With the transfer of data on diskettes the following specifications are necessary before the transfer is initialed:

(a) Size of diskette (sizes may vary between 3 1/2, 5 1/4 and 8 1/2 inches)

(b) Whether they are single or double sided

(c) Whether they are single or double density

(d) Whether the sectoring is soft or hardware managed

(e) The recording mode, which is depending on the operating system of the hardware (the usual, at the time of writing of this publication are UNIX, MS-DOS, CP/M, IBM-DOS, but new ones are developed constantly, the ALERT HFS operates with a QNIX).

Standard format

Where data are to be transferred from or to a number of different agencies, it may be worthwhile for a standard format to be developed in which data would be written for transfer. By the adoption of such a standard, the programming commitment involved in writing and reading such transfers would be reduced. Also, standardization of the format would tend to encourage the standardization of data presentation and analysis.

In connexion with the question of the definition of a standard transfer format, insufficient experience has been obtained to allow all the implications of the procedure to be identified. However, any such format should meet the following guidelines:

(a) Be oriented towards 80-character records;

(b) Allow for the provision of a character dictionary to assist decoding;

(c) Contain header information specifying the identification and location of the data source, together with data and time period specification;

(d) Allow for the inclusion of plain language comments within the file for the inclusion of quality and instrumentation remarks;

(e) Provide for a data field which contains sufficient significant figures and which can be decoded by a fixed format;

(f) Allow for the identification of the data fields, so that time series data can be consolidated to remove redundant items, e.g., time series data can be formatted as a sequence of data and time entries;

(g) Allow quality indications to be associated with particular data items;

(h) Provide positive indicators of data sequence start and finish;

(i) Provide enough information to allow an independent check that all data of a sequence are present.

Creation of HFS files

The data base management sub-system of any HFS system depends on many conditions: those of the hydrological nature of the basin, type of forecasting procedure, type of available hardware which can range from simple desk calculators, passing through microcomputers (Apple, Tandy, Rainbow, Cromemco, IBM PC, PCXT, PCAT, etc.), ending with medium (mini) and main-frame computers (PRIME, VAX, BOURROUGHS, IBM and others). As a rule this sub-system absorbs the computing operation of the model development and model execution (operational forecasting) program, as well as updating and often even data bank creation programs. The files created can be of many types, the most common are:

- Station files;
- Catchment parameter files;
- Model parameter files; and
- Program execution files.

The data base, "the knowledge base" of the HFS, may be created by adaptation of an existing data processing software existing at the service which is in charge of HFS. Some of these procedures are described in detail in the WMO Guide to Hydrological Practices (WMO, 1983) and in the above-mentioned WMO/FAO Guidelines for Computerized Data-Processing in Operational Hydrology and Land and Water Management (WMO, 1985). With the design of an HFS however it is more efficient and indeed in many cases more economical to create a special data base for the HFS, even at the cost of transfering data from a general hydrological or water-resource national or basin data bank to the data base of HFS. Furthermore as the HFS often uses a stand-alone computer, the adaptation of software from the operating system of the general data bank to this of the HFS would probably be more expensive than using a specialized data base management software, already prepared for coupling with the forecasting procedure (model) and transfer the historical data. In this connexion, however, a warning has to be sound. Several ready made, pre-packaged, data base management software do accept only one particular forecasting model, which may not be the best for the river or basin to be forecasted. Thus when selecting such a pre-packaged software the suitability of the model which goes with it has to be ascertained before the software is adopted. As this software is often commercially marketed, the caution must be exercised not to spend funds on a system that will not deliver the service. An example of a pre-packaged data base management software, which is designed to work with the model of the US National Weather Service River Forecasting Center in Sacramento - the "Sacramento model" is given at the end of this section. This software called "ALERT" is of public domain, it may be obtained free of charge through WMO HOMS. The same model is used with several other software packages, including a data base management, commercially marketed by several hardware companies in the US. The prices of these software packages vary, sometimes considerably. For comparison the reader may wish to note that a very large and very good data base management software, coupled to three different models and specially designed for the River Niger in Africa delivered to Niamey with training both in Europe and in Africa of a large group of user's personnel was purchased by WMO for about one hundred thousand US$. The HOMS of WMO contains a complete HFS software package (sequence) including two simple models, developed specially for several rivers in South-East Asia. The data base management of this package was built bearing in mind criteria of simple structure of programs, comprehensive approach to satisfy also possible future demands of the user and avoid redundancy (repetition of functions) within its different components. The data base performs five operations:

(a) initialization (of data files)
(b) addition and deletion (of data in the files)
(c) modification (of data in the files)
(d) selection and retrieval of data
(e) cleaning of data files.

The above operations can be executed either by optional sub-programs integrated in the user's sequence of programs (a menu) or by independent programs reaching into the files without calling upon the user's menu of programs. The software is written in FORTRAN IV.

Selection of a computer for HIS

Depending on the sub-system described above and on the forecasting procedure selected, if the computer to be used for the FORTH operation is not already available, the question arises as to what type of computer is to be chosen. The selection depends on many circumstances, some of which are entirely independent of the technical design and character of the HFS. Nevertheless only the latter can be discussed here. From the experience of HFSs established by WMO, if a computer is to serve <u>only</u> such a system, it is assumed that it will have to perform the following tasks, assuming that the data collection is computerized (Miller, 1982):

- Interrogation of remote gauges to collect real-time data;

- Alert of the forecasting centre (and possibly users) to imminent dangerous developments in the hydrological behaviour of the basin;

- Data checking and validation (error detection);

- Primary data processing (e.g., stage-discharge conversion);

- Archiving of data in a form acceptable by the data bank and external data processing system (for possible publication);

- Storing and running the program for the development (calibration) of the forecasting procedure (model);

- Storing and running of the operational (real-time operated) forecasting procedure (model) programs;

- Display of forecasting results, possibly for their transmission to remote users;

- Updating of both forecasts and of the parameters of models;

- Running of operating system (main program) controlling all above sub-systems.

In calculating the size of the CPU and of peripherals one can establish the data flow and processing, compute the space requirements for each and proceed from there. However, with the rapid development of micro-computers, their transition to 16 and even 32 bits/word usage, and overlay possibilities, the size of the CPU becomes a secondary concern in the selection of the computer. The US NWS Forecasting Center in Sacramento (the home of the Sacramento model) considers the following criteria to be evaluated in the selection of a computer for a HFS (Burnash, 1982):

- Core size

(a) addressable words
(b) word length (in bits)

- CPU speed

- Communication ports

(a) serial
(b) parallel
(c) number

- Expandability

(a) core
(b) communication ports

- Mass storage capability (discs or tapes)

(a) size (in words)
(b) access speed

- Diskette capability (e.g. tape interface, etc.)

- Service availability

- Type of operating system

- Real Time Clock availability

- Interrupt servicing

- Hydrological software availability

These factors cannot be simply evaluated as they interact with one another in a manner which requires knowledge of the concrete problem and its relationship to the above factors to obtain a valid analysis. Considerations which are relevant to these factors include:

- Core size is not directly related to overall program size, the overlay potential of a system with a good interrupt structure, high speed disk and an effective operating system can reduce core requirements to less than ten per cent of that required by a system without such capabilities. The apparent program size and on-line file requirements do not necessarily determine core size, an effective combination of other factors may provide a more cost-effective solution. Associated with core size is word length. The size of addressable core is related to word length (generally 8, 16, or 32 bits). With today's microcomputers the choice of word size is generally economic rather than technical.

- CPU speed - other things being equal may be significant - however, the speed of new microcomputers leaves little to be desired. The current technology in widest uses chips with 16 and 32 bytes. Microcomputers with this new technology have computing capabilities which may exceed that of many of the IBM 370 models. The primary limitation in either case is likely to be associated with the disk accessing. Disk accessing is related to the operating system and the software design of the data base and the application programme.

- Communication ports are one of the few criteria for which hard guidelines can be developed. An analysis of the incoming and outgoing lines needed for communications, peripherals, and terminals and their serial or parallel characteristics allows a determination of the required configuration. Unfortunately, some computers do not provide a sufficient interrupt possibilities to service all equipment which may be physically connected to the many ports which are competing for resources. Always analyze the ports which are essential to the programme vs. those which are desirable. Under some circumstances, a single low priority terminal can double the system cost.

- Expandability: In the event that phased acquisition is being considered, the ability to add terminals, core, and peripherals may be significant. Verify that required upgrades are possible without an operating system change which would require a totally new software package.

- Mass storage requirements involve not only size, they also involve access time. The mass storage requirement is a function of core size, data base requirements, the operating system and the software design of application programs. The data base design and supporting software may affect storage requirements by more than an order of magnitude and can have similar impacts on access time. As an example, special algorithms allow a sensor value, the month, day, hour and minute (to within 3 minutes), all to be stored in two integer words. The impacts on the size of file storage, required data access times, core size for data arrays, and cost effectiveness are all negatively affected by a failure to take advantage of custom software.

- Diskette capability is highly desirable for software and data exchange. Speed of access is generally a convenience, not a necessity. The size of diskette (storage potential) may be important as a convenience, but size limitations can generally be overcome with software when a hard disk is available.

- Availability of service: This factor is essential for any operational unit and may preclude any other decision.

- Operating system: Is it real time? Does it have high level language capability? Is high-level language code compiled to machine code, intermediate code (such as P-code) or interpreted?

- Real time clock capability is essential in data systems. Is the clock addressable by higher level language? Is special software writing required? Can the clock generate interrupts? Does the clock keep time during interrupts? (Some computers have problems.)

- Interrupt servicing: How many lines can be active? Will heavy interrupt use generate crashes?

- What software is already available? What are the reprogramming costs?

- Graphics facilities (is VDU graphical, what is its resolution, number of colours/shadows intensities, is it possible to plot graphs on line printers without digital plotter, is the software support readily available?

Computer selection is therefore a system of trade-offs, the state-of-the-art for various factors is changing so rapidly that guidelines become archaic as soon as they are written. A continuing re-analysis is required. The most cost-effective process consists of maintaining a basic position for software development and shifting to an updated hardware configuration only when software growth or service requirements demand it. Under current conditions this can impose a substantial software redesign on a three or four year cycle. As a basic guideline for new systems which can perform at the current level of hydrological data collection and analysis capability, a hydrological microcomputer system consisting of:

- 256K core;
- High speed CPU;
- Four serial and one parallel communication ports;
- Core and communications expansion capability;
- A hard disk drive;
- Floppy disks (diskettes)
- Easily available and rapid service;
- UNIX-like, DOS (MS or IBM) type real-time operating system with FORTRAN and BASIC or PASCAL, at least;
- Real-time clock; and
- Interrupt capability,
- Networking capability or some commonly used terminal emulation software availability,

A computer of this type, which could be recommended for any significant project, will range upward from US$5,000 (five thousand dollars) for the basic unit (without optional peripherals) with selection being substantially determined by service availability and the availability of custom software to meet project requirements.

At all times, it must be remembered that:

- A computer system is of no value if it cannot be maintained in an operational status (i.e., availability of service);

- The cost of software transfers to another computer almost invariably introduces costs which could be in some cases larger than the cost of the computer system - this is particularly true of microcomputer systems and becomes a very major factor if the original code was not prepared in a high level language (i.e., includes assembler routines with special handling of I/O ports).

The experience of WMO demonstrates that in developing countries the most important criterion is the availability of service through a local dealer of the computer, or easy transport to the nearest servicing point. Administrative considerations and availability of similar computers in other institutions (non-hydrological ones as well) have also to be considered in these countries.

Pre-packaged data base management with model designed for specific micro-computer type (ALERT)

As indicated above, in the US exists an HFS which is delivered with all sub-systems pre-packaged, to the extent of feasibility. The system's "ALERT" software in its non-commercial version, without the hardware, is included in the HOMS of WMO and can be obtained free of charge on a diskette. The hardware (hydrological and meteorological sensors, telecommunication and equipment and the micro-computer) can be purchased also as a package from at least three US companies. The micro-computer recommended by the manufacturers of the HFS is an IBM-XT with a QNIX operating system.

The ALERT XT HFS collects event data from a variety of sensor types. It files this information in a data base that allows user access for data display and analysis. Alarm criteria can be set for each sensor in the system. The ALERT program checks if the alarm criteria has been exceeded each time a report from the sensor is received and performs an alarm action if so. The alarm action can be selected from those provided with the software (flashing screen, audible alarm) or user created actions can be performed.

Besides the ALERT data collection, filing and alarm processes, the system provides several methods of data display. Groups of sensors can be displayed for the last 24 hours or any specified past 24-hour period. Single sensor displays can be made for any number of reports in any desired time period. The data display software includes a data filter routine that removes reports that are not continuous with other reports in the data file.

Utility programs are provided to establish sensor reports to be collected and filed in the database. Sensor names are definable up to 20 characters. Water level sensors can be calibrated to observed stages. Data base review programs are also provided for system review and maintenance.

A data archive program is provided to create monthly archive files of system data in text form.

The system provides an easy man-supervised control by allowing display of data acquired by individual sensors as indicated in the following example:

National Weather Service Hydrologic Data Collection System

Date 6/29/84 Time 9:15

Station identification

	# 221	# 222	# 224	# 217	# 217	# 220
Data						
Last 10 min	0.31	69.0	1010.0	19.9	154.0	0.00
Last 30 min	0.31	69.0	1010.0	19.9	154.0	0.00
Last 60 min	0.31	69.0	1010.0	19.9	154.0	0.00
For 24 hours						
Ending at 900	0.33	68.0	1009.3	18.1	171.0	0.00
Since 900	0.31	69.0	1010.0	19.9	154.0	0.00
800 Thru 900	0.33	68.0	1009.3	18.1	171.0	0.00
700 Thru 800	0.38	64.0	1009.3	20.1	154.0	0.00
600 Thru 700	0.39	62.0	1008.3	14.6	160.0	0.00
500 Thru 600	0.44	59.0	1007.3	18.9	154.0	0.00
400 Thru 500	0.43	59.0	1007.3	18.0	143.0	0.00
Since 500	0.31	69.0	1010.0	19.9	154.0	0.00
2300 Thru 500	0.43	59.0	1007.3	18.0	143.0	0.00
1700 Thru 2300	0.36	68.0	1006.7	19.9	149.0	0.00
1100 Thru 1700	0.18	95.0	1005.0	18.7	188.0	0.00
500 Thru 1100	0.27	83.0	1006.7	16.6	182.0	0.00

The ALERT system will also do alarm criteria checking for any sensor defined in the data base. The criteria checking occurs each time a sensor report is received and filed in the data base. A sensor report must pass a continuity check before any criteria checking is done. When a sensor's data file is examined for criteria checking, the data is first filtered to exclude any discontinuous reports.

The current alarm criteria are:

- value to be equalled or exceeded
- value to be equalled or fallen below
- positive rate of change in a time period, with an optional threshold value which must be equalled or exceeded before the rate of change criteria applies
- negative rate of change in a time period, with an optional threshold value which must be equalled or fallen below before the rate of change criteria applies.

The alarm criteria are set in the units defined for the sensor. For the rate of change criteria, the sensor data record is examined for the specified time period and the change between the current report and the oldest report in the period is used to determine the rate.

The ALERT data collection and analysis system requires a fairly large amount of storage space on a computer in order to store its programs and data files. The following is a list of the storage requirements for each part of the system.

Type of file	Storage required
Basic data collection system	
Programs	650,000 bytes
User's manual	140,000 bytes
Data files	316,000 bytes + 30,720 bytes per sensor for wind, temp,humidity, 15,360 bytes per sensor for other types (stream, precip., etc.)
Stream rating tables	520 bytes per rating table
Map data displays	2,000 bytes per map
Alarms	approx. 50 bytes for each active alarm
Streamflow simulation system (model)	
Programs	500,000 bytes
Data files	20,000 bytes per basin

In addition to the amount of space required for each of the above files, there are some files that keep growing with time and should be periodically edited in order to keep them for overloading the hard disk's storage capacity. If you have alarms get on any of your sensors, the following files will grow slowly, but continually.

/user/alert/alarm/alarm. log
Grows by about 70 bytes each time an alarm goes off.

/user/alert/alarm/crit.a
Grows by about 70 bytes each time an alarm is set. (It does not shrink when an alarm is disarmed.)

From the above description it appears that for a smaller HFS the ALERT may be the solution to avoid expensive custom written software for all parts of the functions. The author's only reservation is the inability, for the time being, of this HFS to use other river simulation than the "Sacramento model". As indicated there exists similar standard packages in Europe and particularly within the software developed for WMO HFS projects, which have not such limitations. They have however other limitations, in particular they are not designed for any specific hardware configuration of sensors, telecommunication and computers. As in many similar cases, the designer of the HFS should use the services of a highly specialized consultant to find out the most efficient and economical solution.

2.5 FORECASTING PROCEDURES (MODELS)

Two modes of existence of forecasting procedures

Every forecasting procedure (model) exists in two modes (computational forms):

(a) development mode (calibration)

(b) operational mode (actual forecasting)

The complexity of the first mode may not be relatively equivalent to the second one, in other words, a model which has very complicated and large requirements on computation time and facilities (computer size) in the development (calibration) mode may be operated for the actual issuance of the forecast with very simple mathematical formulation and minimal computational facilities, bordering on manual use of a pocket calculator. Reciprocally, a relatively simple, mathematically formulated model needing less time and relatively smaller computational facilities in development (calibration) may be more difficult to operate for the issuance of an actual forecast, and may require larger computational facilities, in some cases requiring at least an electronic micro or mini-computer.

The above differences in development and operational algorithms may be caused by several factors of which the number of internal parameters of the model and the development (calibration) data needs are considered the most important. It should be noted, however, that the above requirements of and differences between models in both modes do not necessarily influence the final choice of the forecasting procedure, which depends on many other factors, as indicated in section 3.2

Classification of models for hydrological forecasting purposes

It is necessary to stress that the classification of models for hydrological-forecasting purposes is not necessarily identical to the classification of models in general or for other purposes such as computation of design data, extrapolation of records, research purposes, etc. Some models, while based on stochastic principles or using exclusively statistical approaches, may function as deterministic models when used for real-time forecasting. In this connexion, it should be stressed that any model used for real-time forecasting is, by definition of the latter, a deterministic model.

As pointed out by Gabris (1985) hydrological systems have an imminent quality of predictability. This is resulting from the time lag between the cause and effect, or between the input and output in these systems. Despite of this imminence not all hydrological models incorporate this lag, and those which do not are not able to forecast. Thus in the following text we will limit ourselves to such models which are able to produce forecasts in real time.

The WMO project on intercomparison of conceptual models for operational hydrological forecasting used the following two classifications of conceptual models, which were not mutually exclusive, and thus could be used simultaneously to one another:

(a) (i) Models of basic basin response (these models are not able to simulate effects of human interference on the behaviour of the basin);

(ii) Models of the response of a system of rivers, including human interference (these models are able to simulate short-term effects of human interference on the behaviour of the basin);

(b) (i) Models involving volume of runoff;

(ii) Models involving distribution of runoff;

(iii) Models involving the routing of flow through the channel system.

Forecasting with the aid of conceptual models generally involves all three of the components under (b); but on occasions only one or two are used, e.g. (i) and (ii), or just (iii).

The project further classified the tested models in category (b) (i) above into those displaying the following characteristics:

(a) explicit moisture accounting

(b) implicit moisture accounting

(c) systems approach with the use of indices (of soil moisture).

This classification is, of course, not all-encompassing, as only ten conceptual models were tested in this WMO project. A survey of HFSs in Europe (see Annex 2a) indicated that many other models are used, with the majority being more empirically based (e.g. correlations) than using theoretical concepts (e.g. conceptual models). The author of this text therefore proposes the following classification of models used for HFS, which he considers practical and based on actual operational HFS in the world.

A. *Purely deterministic forecasting procedures (models)*

A.1 Hydrometric data-based (involving only streamflow processes)

A.1.a Correlations of stages and/or volumes (discharges)

A.1.b System approach to streamflow (hydrologic routing)

A.1.c Hydraulic routing using

(i) Dynamic wave
(ii) Diffusion analogy
(iii) Kinematic wave

A.2 Hydrometeorological and hydrometric data based models (involving rainfall/runoff and streamflow processes)

A.2.a Correlations using physical variables and parameters or indexes (such as API)

A.2.b System approach to the basic basin response to rainfall

A.2.c Distributed parameter approaches (hydrological and hydraulic)

A.2.d Conceptual moisture accounting using
(i) soil moisture indexes
(ii) implicit moisture accounting
(iii) explicit moisture accounting.

B. *Hybrid stochastic-deterministic forecasting procedures (models)*

B.1 Using only time-series stochastic parameterization.

B.2 Using a system approach to the basic basin response and time-series stochastic parameterization.

It should be noted that often some models of the categories A.1 and A.2 are combined in one single forecasting procedure and that lately categories A and B are being combined in the so-called "self-tuning algorithms", namely models of category A are updated in operational mode (see above) by procedures of category B.

Principles and mathematical formulations of forecasting procedures (models) most frequently used

The above classification includes a total of ten sub-categories in category A and two sub-categories in category B. The above-mentioned WMO survey of HFSs in Europe (see Annex 2) indicated that in 17 countries a total of 74 forecasting procedures are used of which 35 are correlations (A.1.a, A.2.a or B), 16 are routing procedures (A.1.b, A.1.c and A.1.d) and 23 are of another nature, mainly falling into conceptual moisture accounting (A.2.d) or system approach (A.2.b) categories. The real world is therefore and understandably even more varied than the general classification given.

A.1 *Hydrometric data-based models*

A.1.a *Correlations of stages and/or volumes (discharges)*

The principle of category A.1.a can be best illustrated by Figure 7. The corresponding points of hydrographs of discharges of the three stream gauging stations, in particular the peaks of the floods, can be correlated for both discharge and time of propagation (lag time) from upstream to downstream stations, providing that a sufficient number of observed hydrographs exists at all stations. Knowing the discharge and its time of occurrence at the upstream station, such correlation then yields statistically the most probable discharge at the downstream stations (one, two or several, depending on the availability of data). Such a correlation may be increased in accuracy by including in it additional variables (multiple correlation) such as inflow from the lateral catchment between upstream and downstream stations. Such inflow may be based on some empirical approach to runoff estimation from rainfall or simply using the rainfall itself as an additional correlation variable.

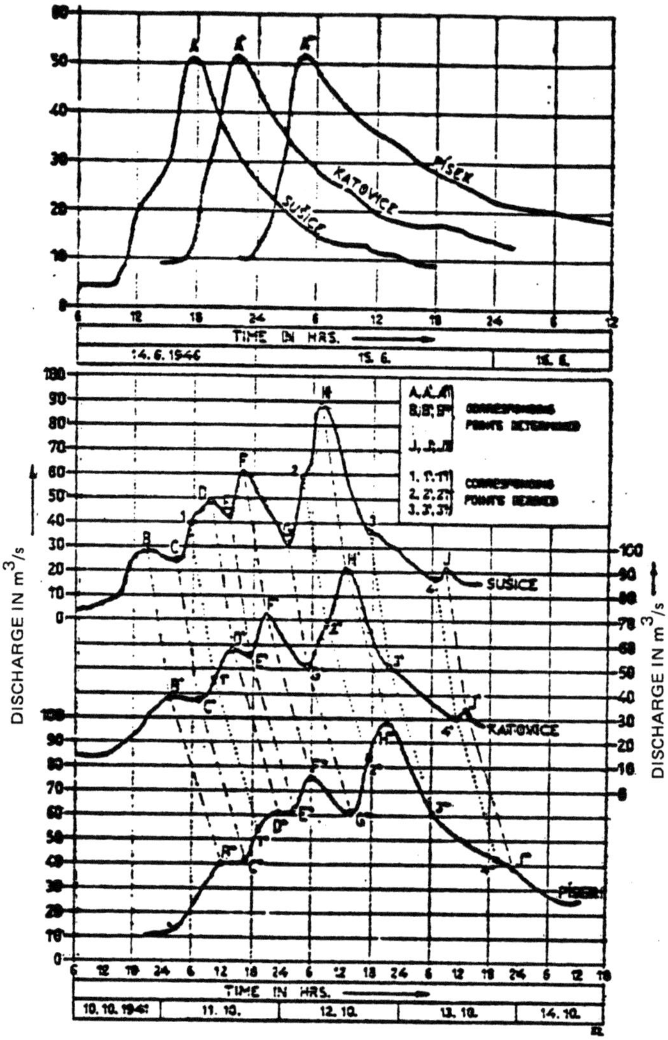

Determination of corresponding characteristic points
(Otava River)

Figure 7. Forecast using corresponding characteristic points.

A.1.b *Hydrological routing*

The principle of hydrological routing in models of A.1.b type (sometimes interpreted as a system approach to the river reach) is based on the storage equation

$$O = I \pm \frac{ds}{dt}$$

where

O = outflow from the reach

I = inflow in the reach

$\frac{ds}{dt}$ = change of storage in the reach.

The above routing principle can be written as a linear differential equation providing that the storage relationship with either the inflow or outflow is linear. This equation represents a linear system of which the parameters can be identified (solved) either as a simple convolution integral or as those of a system of linear reservoirs of Muskingum type (cascade), leading to several forecasting models (Nash, Kalinin-Milyukov, Svoboda, etc.). The theory and practice of this approach is amply documented in most textbooks, and an exhaustive treatment is given by Dooge (1973). An example of this approach is also given below in category A.2.b.

A.1.c *Hydraulic routing*

The hydrological routing is used in cases when there are data only on discharges and corresponding stages (stage-discharge relationship). When data is available on the topography of the channel and its hydraulic characteristics (e.g. roughness), hydraulic routing methods can be used to forecast discharges in downstream stations from upstream observation of discharge. These methods are based on the differential equations of motion and continuity in the channel (Saint Venant equations) and their solution under simplified conditions as indicated in category A.2.c below.

A.2 *Rainfall/runoff models*

In order to increase the lead time of the forecast the use of quantitative information on rainfall and other hydrometeorological data in the basin, together with data on the streamflow, is the principle of rainfall/runoff models.

A.2.a *Multiple correlations*

The simplest models to develop, providing sufficient historical data exist, are based on multiple correlation. They are an expression of the basic equation of water balance (hydrological cycle) with parameters ascertained by multiple correlation of observed input with observed output.

Mathematical formulation:

$$y(t) = a_1X_1(t - \tau_1) + a_2X_2(t - \tau_2) \ldots$$
$$+ a_3p_1(t - \tau_1) + a_4p_2(t - \tau_1) \ldots + e + Cst$$

$y(t)$ = Forecast variable (streamflow) at time t

$X(t - \tau_n)$ = input variable (rain) at time $t - \tau_n$

$p_1, p_2 \ldots p_n$ = parameters of basin at time $t - \tau_n$

e = residual error

Cst = constant

$a_1, a_2, a_3 \ldots a_i$ = partial correlation coefficient

Advantages: simple computation, high reliability and accuracy.

Disadvantages: Each correlation is unique for the forecast variable (best predictors differ from variable to variable and basin to basin); no insight in the process, no improvement possible in the structure of this type of model. Same models can be used with graphical (co-axial) correlation (US WB API model). Multiple correlation model with continuous hydrograph (APIC model) is described by Sittner et al. (1969) as an extension of the original US Weather Bureau's model which enables a continuous hydrograph synthesis to be made. The model consists of four basic parts: a relation for computing groundwater recession, a method for computing the groundwater flow hydrograph as a function of the direct runoff hydrograph, an API-type rainfall-runoff relationship and unitgraph. The rainfall-runoff relationship is of the incremental type, yielding a runoff computation for each hourly period instead of the total storm runoff. This and the inclusion of a new parameter, a retention index, overcome some of the problems with the original model discussed earlier. The procedure being necessarily more complex requires a computer solution.

A.2.b *System approach to the basic basin response to rainfall*

A recently developed Constrained Linear System (CLS) Model can be described as representative of this category. The CLS model has its basis in linear system theory which, in the form of the unit hydrograph, has enjoyed much success in hydrology as a method for rainfall-runoff modelling. As a starting point, CLS assumes that a linear relation exists between an input, U(t), which may be rainfall or an upstream discharge, and an output, Q(t), which is the discharge at a point for which forecasts are to be made. In continuous time, the relationship may be expressed by the well-known convolution integral representation of a linear system

$$Q(t) = \int_0^t v(t-\tau) U(\tau) d\tau,$$

where the function v(t) is the impulse response function (IRF), or instantaneous unit hydrograph (IUH), or kernel function. CLS uses the discrete time form of the above equation,

$$Q(t) = v_1 U_t + v_2 U_{t-1} + \ldots + v_k U_{t-k+1},$$

where Q_1, Q_2,... and U_1, U_1, U_2, ... are the discharges and observed inputs at fixed time intervals, and v_1, v_2, ..., v_k are the ordinates of an impulse response function with a time base of (k-1) time units. Thus flow at time t is predicted simply as a weighted combination of k present and past inputs, U_t, U_{t-1}, ..., U_{t-k+1}, the weights being given by the ordinates v_1, v_2, ..., v_k.

The estimation of the v weights is essentially that of deriving the discrete form of the instantaneous unit hydrograph (IUH). The essential element of CLS, as developed by Natale and Todini (1977), is the introduction of constraints into the parameter estimation procedure which help in obtaining reliable and physically realistic impulse response functions (i.e. unit hydrographs). The constrained estimator is less sensitive than the classical estimators (e.g. least squares) to errors in the data, and its greater efficiency means that reliable estimates may be obtained even from small data sets. The constraints imposed are deduced for the physics of the hydrological system, and are such as to preserve continuity and to ensure that flow occurs only in a downstream direction. To preserve continuity, an equality constraint is imposed such that the sum of the $\underline{v}$ weights sum to unity:

$$v_1 + v_2 + v_3 + \ldots + v_k = 1$$

To ensure flow occurs only in a downstream direction, the inequality constraint that the $\underline{v}$ weights should be zero or positive is imposed:

$$v_i > 0 \qquad , i = 1, 2, \ldots, k.$$

The equality constraint is used with advantage to estimate the runoff coefficient using summational notation

$$\sum_{i=1}^{N} \sum_{j=1}^{k} v_{(i-1)k+j} \sum_{t=1}^{m-j+1} P_{i,t} = \sum_{t=1}^{m} Q_t$$

where $P_{i,t} = U_{i,t}$ are precipitation inputs.

The runoff coefficient ϕ_i for the i-th precipitation input $P_{i,t}$ is

$$\phi_i = \sum_{j=1}^{k} v_{(i-1)k+j} \qquad i = 1, 2, \ldots N$$

In this way, when using the CLS approach, it is not necessary to proceed with net rainfall computation and hydrograph separation prior to estimating the IUH, since the model calculates automatically the runoff coefficient directly from total precipitation inputs and corresponding observed runoffs for each time increment of the IUH. This astuteness of CLS resulting in rapid parameter identification and the possibility of its simultaneous use in routing, makes it more suitable for forecasting in real time.

A further extension of this model is the CLS Non-Linear Threshold Model combined with the API approach, which is a combination of categories A.2.a and A.2.b. The amount of rainfall which will contribute to the storm hydrograph is strongly influenced by the basin soil-moisture conditions prevailing at the time, and this is viewed as the major source of non-linearity in the relation between rainfall and runoff. In order to accommodate such non-linear effects in the CLS model, Todini and Wallis (1977) introduced a threshold mechanism whereby the rainfall was split into two sequences on the basis of a basin wetness threshold criterion. In its current form the threshold mechanism employs the antecedent precipitation index API_t computed at time t,

$$API_t = K.API_{t-1} + P_t,$$

to split the rainfall P_t into two sequences. If T denotes the API threshold value at which the split is made, then the two split inputs are defined by

(1) $API_t > T \quad P_{1,t} = 0$

$\qquad P_{2,t} = P_t$

(2) $API_t \leq T \quad P_{1,t} = P_t$

$\qquad P_{2,t} = 0$

Each split input sequence is then directed through its own linear system, one having a fast response ($P_{2,t}$) and the other a slow response ($P_{1,t}$), with flow at any time t being given by the sum of the outputs from the two linear systems (Figure 8).

To accommodate a seasonally varying API_t constant, K_t, a seasonal API may also be defined as

$$API_t = K_t \, API_{t-1} + P_t$$

where

$$K_t = \bar{K} + \alpha \cos (2\pi \, (t+\phi)/365).$$

A.2.c *Distributed parameter approaches*

In category A.2.c the representative examples are hydraulic models of overland and channel flow. Their definition is: Expression of the runoff process on the surface and in channel systems of the basin by differential equations of motion (Saint Venant) and continuity (Beven, 1985).

Basic equation:

$$g \frac{\partial z}{\partial x} + \frac{\partial u}{\partial t} + \frac{\partial u}{\partial x} + g \frac{Q^2}{k^2} + \frac{q}{F} = 0$$

$$\frac{\partial Q}{\partial x} + \frac{\partial F}{\partial t} = q(x,t)$$

where x = length of flow
z = depth of flow
u = average velocity of flow
Q = discharge
F = cross-section of flow
q = lateral inflow for unit of length and unit of time
$z = f_1(F,x)$, $k = f_2(F,x)$, g = gravity

Analytical solution not available.

Transformations:

(a) Overland flow (addition of terms for velocity and momentum changes resulting from raindrop impact);

(b) Channel system (addition of terms for velocity and momentum changes resulting from influence of lateral inflow into channel and of tributaries at junction-flow points).

(a) CLS Threshold Model

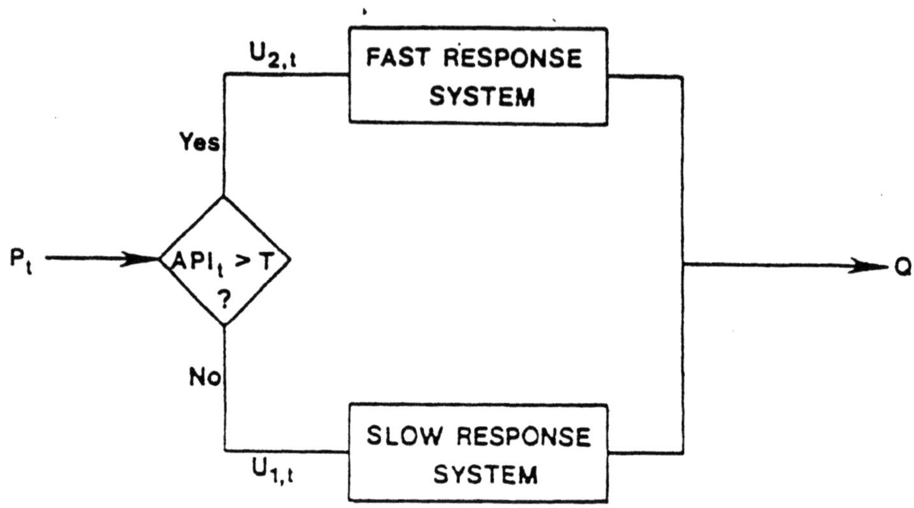

(b) CLS Threshold impulse response functions

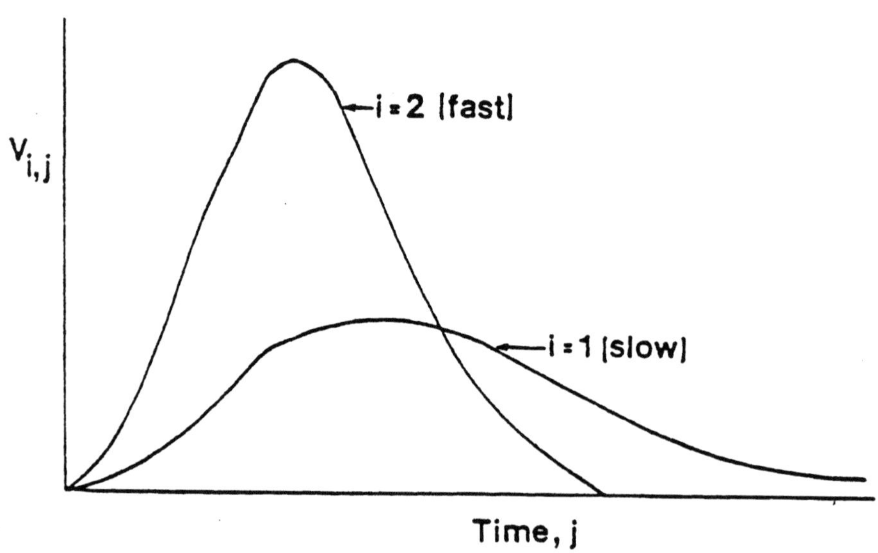

Figure 8. CLS Threshold model.

Numerical solutions available for simplified equations for particular cases:

(a) **Linearizing of the equation for non-steady flow as a dynamic wave (gravity forces prevail, friction forces negligible, inertia forces negligible);**

(b) **Transformation to steady flow (equilibrium of gravity and friction forces) as kinematic wave (discharge in each point unequivocally related to depth and cross-section of flow).**

Resulting equation:

$Q = k\sqrt{i}$ (Manning, Chézy)

$$\frac{\partial Q}{\partial t} + \frac{\partial Q}{\partial F} \cdot \frac{\partial Q}{\partial x} = 0$$

K = discharge coefficient, i = slope, others as above.

A.2.d *Moisture accounting models*

Conceptual moisture accounting models are all in this category. Their definition is: representing different hydrological processes (components) in the basin with simple mathematical expressions, distinction between production function (precipitation minus evapotranspiration and infiltration) and routing function.

Usual input: Precipitation (liquid or solid), meteorological elements influencing evapotranspiration (ETP) at potential rate (air temperature, input to Penman approach, pan evaporation, etc.).

Usual output: Runoff (direct runoff and baseflow), actual ETP, actual soil moisture.

Variable: Physical quantity of either input or output.

Parameter: Constant value associated with a particular basin. Either part of input (taken directly or estimated from physiographical and climatic basin characteristics) or produced by the model as a function of the model's components. The complexity of the model is proportional to the number of parameters. Identification of parameters by calibration (fitting). Possible use of objective function and automatic optimization of the parameters.

Example of use of Moisture Index

SSARR model. Flow chart: two-key points - SMI (Soil Moisture Index) and BII (Baseflow Index). Three runoff volumes - baseflow, sub-surface flow and surface flow independently routed into a

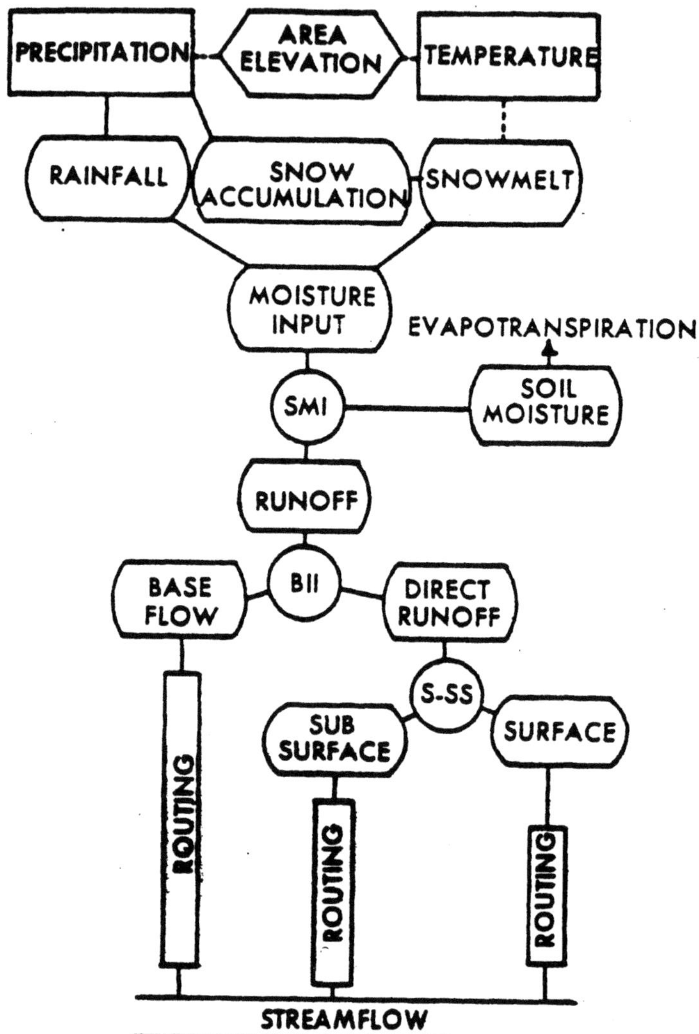

Figure 9. SSARR Watershed model.

hydrograph through a variable number of storages. Several subjective decisions necessary, although optimization procedure available for some parameters. (See Figure 9).

The model combines basin runoff techniques producing surface and subsurface flow with river and reservoir routing techniques. Special optional routines include snow accumulation and snowmelt, soil moisture, and solar radiation.

The model uses 12 to 18 basin coefficients and three river coefficients. If all optional routines are used, 25 to 30 coefficients would be required. Five to ten years of simultaneous meteorological and streamflow records are recommended for calibration.

Inputs: Daily data from multiple stations include observed precipitation and streamflow, plus forecast precipitation and maximum air temperature. Snow conditions, soil moisture, solar radiation forecast, etc. can either be input or allowed to be computed as functions of other parameters.

Output: Surface and sub-surface flow, percentage runoff, snow water equivalent, snowline elevation, soil moisture, river stages and discharges, and reservoir inflows, outflows and storage volumes.

Operational requirements and restrictions: Installation and modification of the program needs professional personnel with specialized training. The program is written in FORTRAN and is designed to operate on a medium-sized computer system which includes disc storage units. The smallest system which has been used is an IBM 360/40 WITH 64K bytes of core storage and a random access disc storage drive. A less extensive version of SSARR developed for small to medium-sized computers is called COSSARR.

Example of use of implicit moisture accounting

Tank model - Sugawara, Oaki, Watanabe, Katsuyama (National Research Center for Disaster Prevention, Tokyo) (Fig. 10).

Principle of tank (reservoir) - tanks in series (Top tank surface storage and overland flow, second tank hypodermic (intermediate) runoff, third and fourth groundwater and baseflow). The number of side outlets is arbitrary. If several side outlets, relationship between the storage x and discharge y is not linear (either represented by broken line or curve). Infiltration I proportional to x. But it increases only to a saturation point (H_s) when I becomes I_s = H_s = constant. This principle is valid with the assumption that there is always a moisture supply to all the tanks (humid regions). A composite tank model was devised for basins with dry spells. Basin divided in zones departing from the river (zones are not isochrones but so-moisture zones - wettest

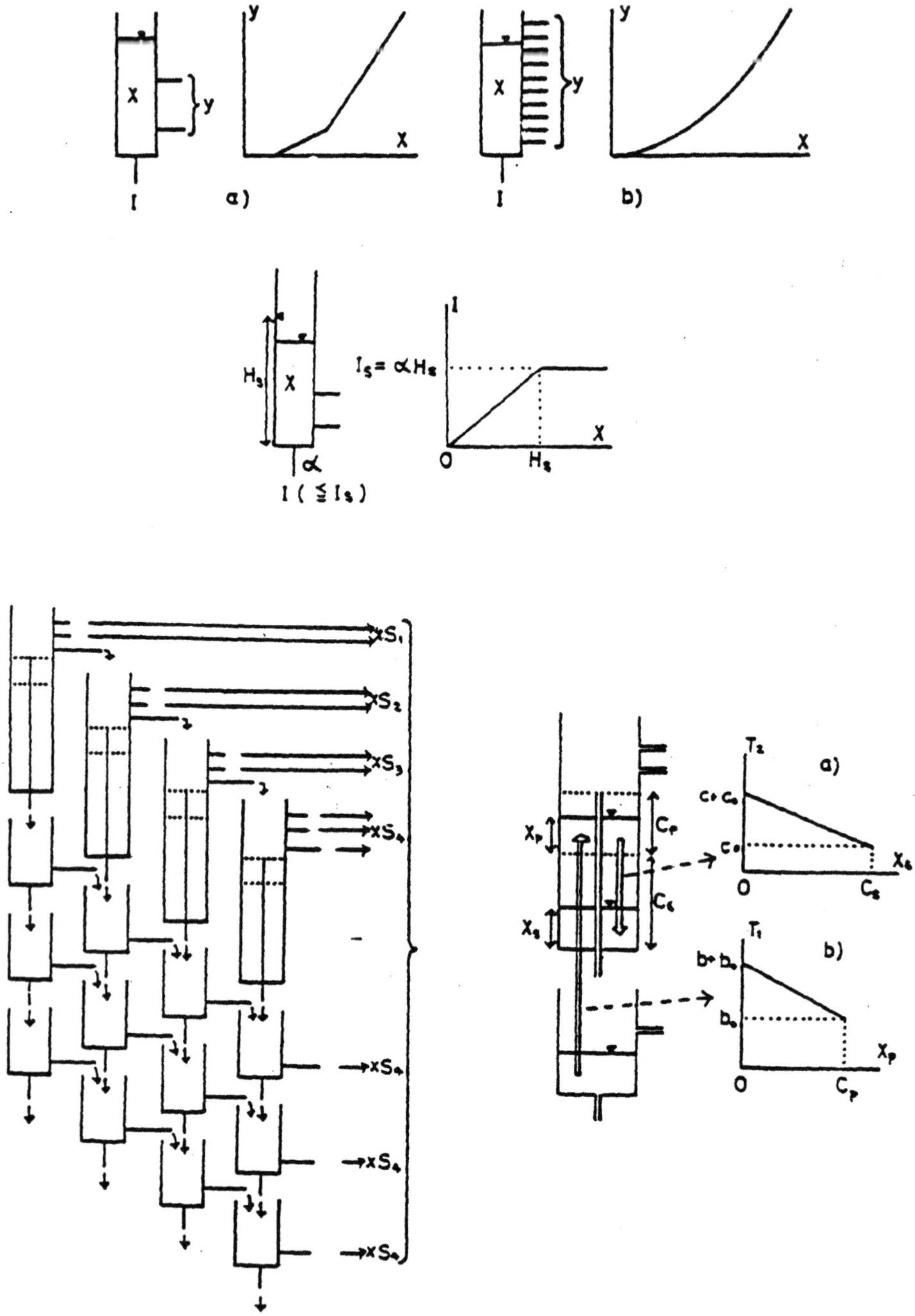

Figure 10. Tank model.

next to river). The top tank in each zone simulates soil moisture. Two types of moisture are considered - primary and secondary. Movement is in both directions following equations:

$T_2 = C_o + c(1 - x_s/C_s)$ T_1 = velocity of capillary action

$T_1 = b_o + b(1 - X_p/C_p)$ T_2 = infiltration action

C_o, C, b_o, b = Constants

X_s, X_p = Storage of moisture

C_s, C_p = Saturation capacity of soil.

Evapotranspiration: simple constant 0,5 or 0,8 of potential one (0,6 from lower tanks if there is no water in the top tank, 0,8 from top tank with water). Also a constant value (e.g. 6 mm/day). Routing: Route and lag system (first order lag system $1/(S + a)$ where the time a is ascertained by trial and error.

Example of use of explicit moisture accounting

HBV-2 (Swedish Hydrological and Meteorological Service) (Figs. 11 and 12).

Three storages - Soil Moisture storage (No runoff) (SM)

Upper Zone storage (Q_1) (UZ)

Lower Zone storage (GW)(Q_2) (LZ)

Evapotranspiration at potential rate or as function of SM content. Routing component: area time curve. Accounting time interval: 1 day. Data input: Precipitation spatially lumped over the basin at 24-hour intervals. Potential ETP either from Penman's formula (total radiation, air temperature, air humidity, wind run, duration of sunshine. For total radiation albedo of the basin is to be estimated) or from evaporimetric pan. Calibration of model (parameter identification) with runoff input. UZ and LZ conceived as reservoirs. The model has the following eight parameters:

FC, BETA and LP = used for soil moisture-evaporation routine.

K_1, k_2 = storage-discharge constants from UZ and LZ

PERC = recharge rate of LZ

T = timing parameter

p = part of catchment which receives full precipitation and evaporates at pot. ETP.

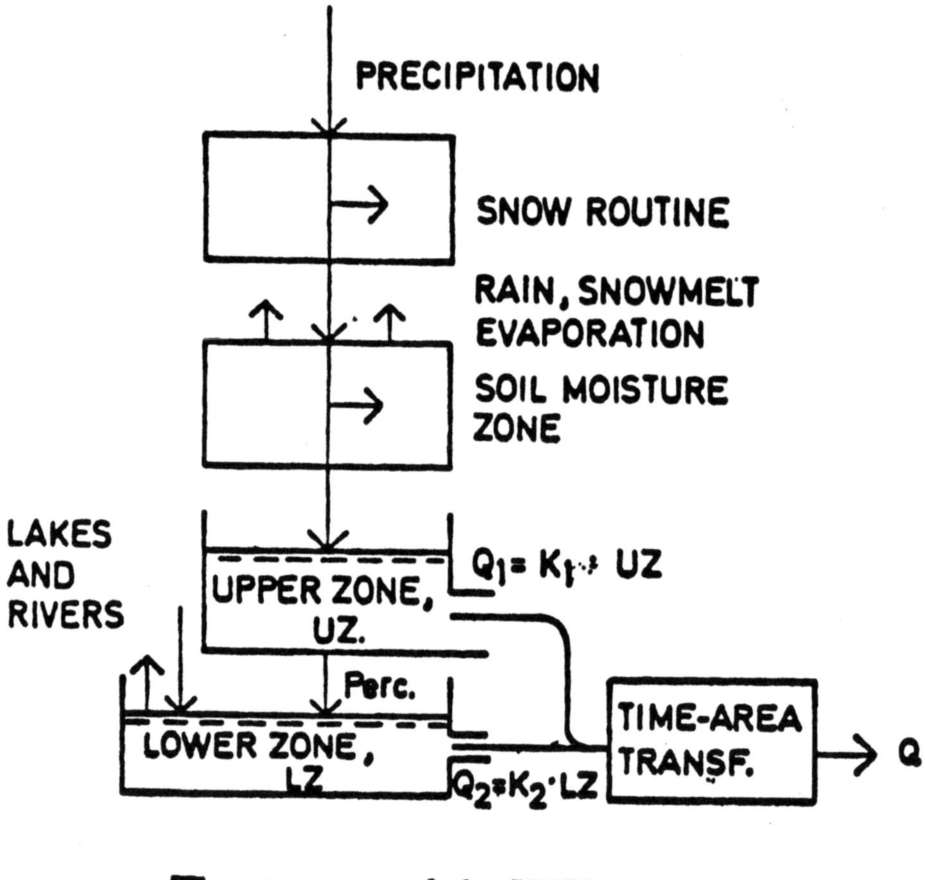

The structure of the HBV-2 model.

Figure 11. HBV model.

SEPARATION BETWEEN SOIL MOISTURE STORAGE
AND UPPER ZONE

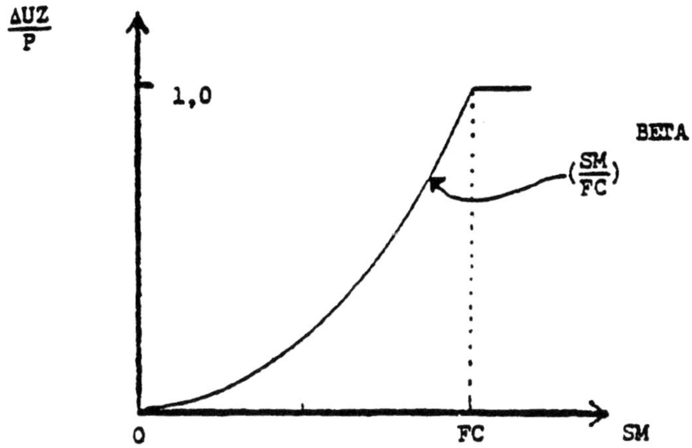

ESTIMATION OF ACTUAL EVAPORATION FROM POTENTIAL
EVAPORATION

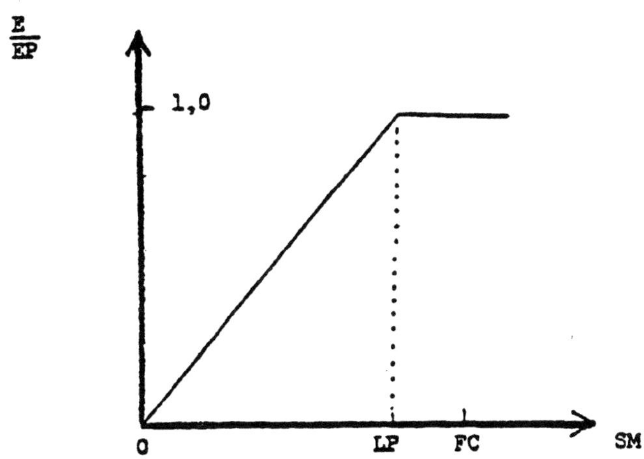

ΔUZ = CONTRIBUTION TO UPPER ZONE STORAGE
P = PRECIPITATION
SM = ANALOGOUS TO THE SOIL MOISTURE CONTENT EXCEEDING WILTING POINT
FC = ANALOGOUS TO THE DIFFERENCE BETWEEN FIELD CAPACITY AND WILTING POINT
BETA = PARAMETER TO OPTIMIZE
E = ACTUAL EVAPORATION
EP = POTENTIAL EVAPORATION
LP = LIMIT FOR POTENTIAL EVAPORATION

Figure 12. HBV model.

Objective function used is $F_2 = (q_{obs} - q_{comp})^2$.

SM = FC, UZ and LZ adjusted to the discharge of the first day.

More complex models use predominantly explicit moisture accounting: Stanford, as adapted by many users. Increase in number of parameters - need for additional data. Similar more complex model, exclusively developed for HFSs, is the Sacramento model (SRECM) developed by the US NWS River Forecasting Center in Sacramento (California).

Structure: A schematic diagram of the model is given in Figure 13. The soil mantle is divided into two zones, upper and lower. Each contains both tension and free water. In the lower zone, there are two types of free water, primary and supplemental.

Functions: Runoff, percolation and drainage.

Direct runoff: As shown, this is an initial abstraction from the rainfall.

Surface runoff: This comes from above the upper zone free water level when the precipitation rate exceeds interflow plus percolation.

Interflow: Shown as being a lateral drainage from upper zone free water.

Percolation: Drainage of free water from the upper to the lower zone. This goes to tension water (left-hand arrow) until tension water is full and any excess then becomes free water. But a portion of percolated water always becomes free water (right-hand arrow) regardless of the tension water deficiency. This is because the physical and configuration of the lower zone varies over the catchment area and there are always some places where tension water is full even though the average condition over the catchment may indicate a deficiency.

Base flow: The two lower zone free water storages represent the slow draining recession (primary) which provides base flow during extended dry periods, and the faster-acting (supplemental) which is active during wet periods. The two functions combined can duplicate the complex recessions observed in nature.

Subsurface outflow: This represents the deep seepage losses which cause some of the flow to enter the river channel, or the sea, at a point downstream from the catchment outlet.

Evapotranspiration: From the upper zone - As shown in Figure 13, evapotranspiration takes place both from tension and from free water. From the lower zone - Evapotranspiration depletes tension water, but there is no direct loss from free water The model

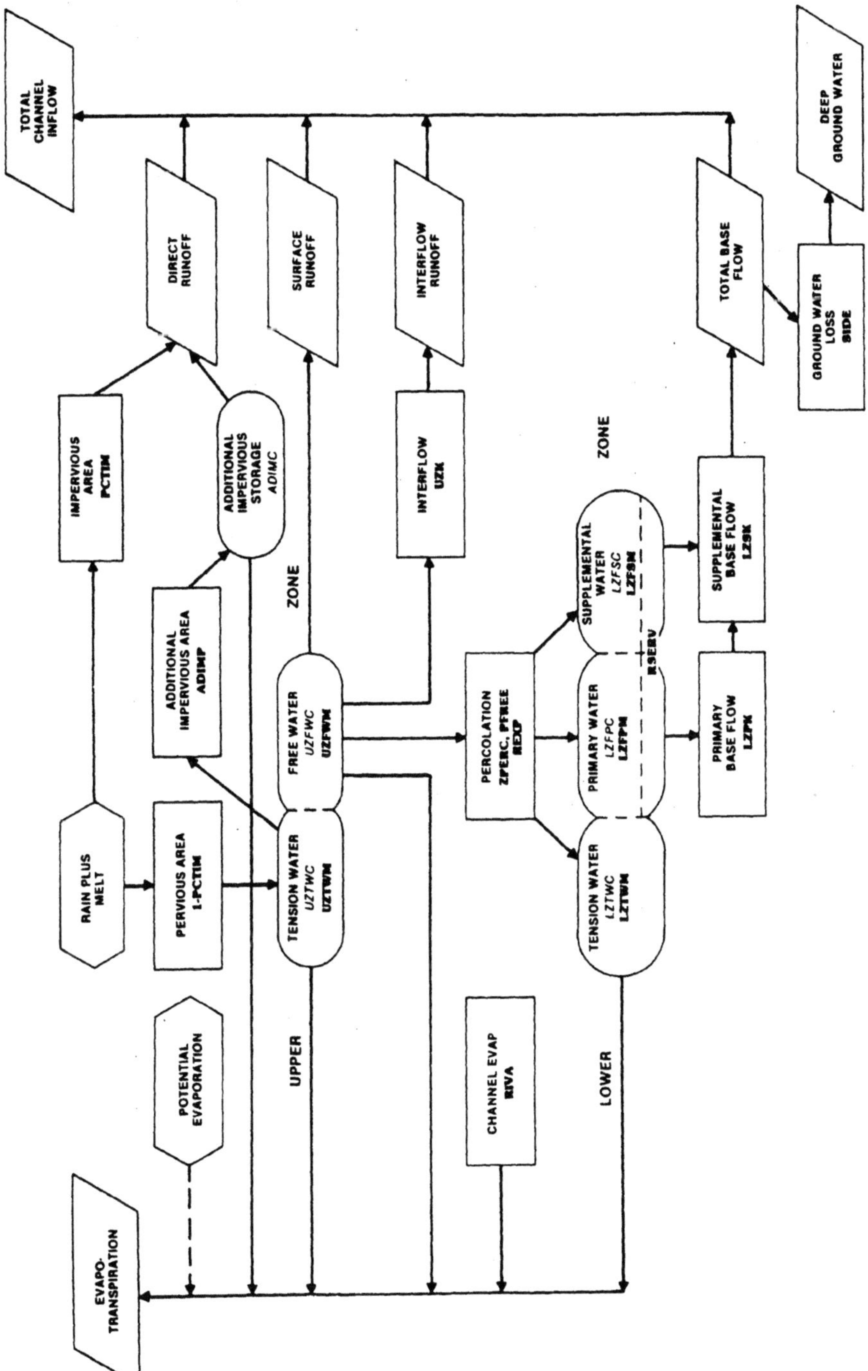

Figure 13. Sacramento model.

provides that, when tension water content is proportionally less than free water, a transfer take place from free to tension (two arrows pointing toward each other just above the dotted lines). The moisture transferred can then be evapotranspired as tension water. Free water storage cannot be totally depleted by such means and the portion labelled "RESERVED" cannot be transferred. From water-covered areas and riparian vegetation - These losses are shown by the two arrows in the lower right corner of Figure 13.

B. *Stochastic-deterministic models*

An example of hybrid stochastic-deterministic models (categories B.1 and B.2) are Transfer Function (TF) models. The basic modelling approach assumes that the observed flow q_t can be considered to be the sum of a linearly deterministic component x_t and a stochastic component z_t:

$$q_t = x_t + z_t .$$

The deterministic component of flow, x_t, is defined so that it is exactly related to the input, u_t (which may be rainfall p_t or some other quantity which influences flow) by the deterministic linear model

$$x_t + a_1 x_{t-1} + \ldots + a_r x_{t-r} = c_o u_{t-b} + c_1 u_{t-b-1} \ldots + c_{s-1} u_{t-b-s-1}$$

where a, c, are parameters, and b is the pure time delay before the flow output responds to a change in the rainfall input. Introducing the backward difference operator, B, defined in $B^b u_t = u_{t-b}$, the above may be expressed concisely in difference equation form as:

$$a(B) x_t = c(B) u_{t-b}$$

where the autoregressive and moving average operators a(B) and c(B) are polynomials in B of degree r and s-1 respectively, i.e.,

$$a(B) = 1 + a_1 B + \ldots + a_r B^r ,$$

$$c(B) = c_o + c_1 B + \ldots + c_{s-1} B^{s-1} .$$

On writing the difference equation as

$$x_t = a^{-1}(B) c(B) B^b u_t = v(B) u_t ,$$

where

$$v(B) = v_o + v_1 B + v_2 B^2 + \ldots ;$$

the series of coefficients v_0, v_1, ... is called the system impulse response function. This is equivalent to the unit hydrograph encountered in the hydrological literature, except that the coefficients are used to define the relationship between total flow and rainfall and are not constrained to sum to unity; in fact their sum is called the gain of the system which may be equated to the runoff coefficient of a catchment. Since in general the order of a polynomial approximation to v(B) will be larger than the sum of the orders of a(B) and c(B), the difference equation offers important advantages by virtue of its parametric efficiency (Box and Jenkins, 1970). Also the dependence of current flow on past flows (that is, the autoregressive nature of the equation) is of particular importance in real-time forecasting applications: it provides a natural mechanism whereby forecasts may be based on the most recently observed values of flow, and not just on past rainfall as in the unit hydrograph (or impulse response function) representation (Moore and O'Connel, 1978).

3. Selection of forecasting procedures

3.1 COMPARISON OF MODELS

In order to help national services to evaluate different forecasting procedures, mainly deterministic conceptual models, WMO conducted, from 1970 to 1975, an intercomparison project of conceptual models (WMO, 1975) for rainfall-runoff modelling and from 1976 to 1983 a project for intercomparison of snowmelt runoff models. A third project, to compare updating procedures and model behaviour in real-time operational conditions, which permit updating and improving of models behaviour is to be conducted in 1987.

The rainfall-runoff models comparison

(a) *The aim of the first project*

(i) To proceed with an evaluation and intercomparison of operational conceptual hydrological models, which use electronic computers to provide short-term forecasts of streamflow;

(ii) To provide information and guidance on the use of such models in various forecasting situations, with regard to specific conditions and accuracy requirements.

It should be emphasized that the intention of the project was not to determine which model fits best in all circumstances.

The project involved the testing of ten operational conceptual hydrological models submitted by seven countries on six standard river catchment data sets from climatologically and geographically varied conditions in six countries. Each data set consisted of two distinct periods: a calibration period (six years), and a verification period (two years) which immediately followed one another. For each data set, the model owners were supplied with the necessary concurrent observed input data (precipitation, evaporation and other meteorological data) and observed output data (streamflow) for the six-year calibration period and only observed input data for the two-year verification period. The observed output data for the two-year verification period were retained by the WMO Secretariat. The ten models as well as the catchments for the six standard data sets are listed in Table IX(a).

(b) *Standard data sets tested on participating models*

For each data set the model owners used the concurrent observed input and output data for the six-year calibration period to calibrate and develop the parameters of their models and employed the additional two years of observed input data in the verification

Table IXa

Standard data sets tested on participating models in the first project

Model name (and abbreviation)	*Name and address of model developer*	*Standard river catchment data sets tested in the project*					
		Bird Creek U.S.A. 2344 km²	*Bikin River U.S.S.R. 13100 km²*	*Wollombi Brook Australia 1580 km²*	*Kizu River Japan 1445 km²*	*Sanaga River Cameroun 131500 km²*	*Nam Mune River Thailand 104000 km²*
Bureau of Meteorology Model (CBM)	Bureau of Meteorology, P.O. Box 1289 K, Melbourne, Vic. 3001, Australia	•	•	•	•		
Girard I Model	ORSTOM, 19, rue Eugène Carrière, 75018 Paris, France	•	•				
Serial Storage Type Model (Tank I)	National Research Centre for Disaster Prevention, 1 Ginza Higasi 6, Chuo-ku, Tokyo, Japan	•	•	•	•	•	•
Serial Storage Type Model (Tank II)	Ditto	•		•			
The Flood Forecasting Model (IMH2-SSVP)	Institute of Meteorology and Hydrology Şos. Bucureşti-Ploieşti 97, Bucarest, Romania	•	•				
Streamflow Synthesis and Reservoir Regulation Model (SSARR)	Corps of Engineers, Portland, Oregon, U.S.A.	•	•	•	•	•	•
National Weather Service Hydrologic Model (NWSH)	National Weather Service, Silver Spring, Maryland, U.S.A.	•		•	•	•	
Sacramento River Forecast Center Hydrologic Model (SRFCH)	National Weather Service River Forecast Center, Sacramento, California U.S.A.	•		•	•	•	
Rainfall Runoff Model of the Hydrometeorological Centre of the U.S.S.R. (HMC)	Hydrometeorological Centre of the U.S.S.R., Bolshevistskaya 13, Moscow 123 376, U.S.S.R.	•	•	•			
Constrained Linear System Model (CLS)	Hydraulic Institute of Pavia University and IBM Pisa Scientific Centre, Italy	○	•	•	•	•	○

Notes: ○ = Test results received in the WMO Secretariat after the conference (1974), that is after the modellers had had the opportunity to acquaint themselves with the data on the output during the two-year verification period of the data sets. These two test results will hereafter be referred to as "delayed test results".

The Kostroma river data set (U.S.S.R.) was tested on the Snowmelt-runoff Model of the Hydrometeorological Centre of the U.S.S.R.

TABLE IXb

Data sets used by participating models

Name of country	Model name (and abbreviation)	River catchment data sets tested in the project					
		Durance River (France)	W3-Watershed (USA)	Dunajec River (Poland)	Dischma Basin (Switzerland)	Illecillewaet River (Canada)	Kultsjon (Sweden)
		2170.00 km^2	8.40 km^2	680.00 km^2	43.30 km^2	1100.00 km^2	1110.00 km^2
Canada	U.B.C. Watershed Model (UBC)	X	X	X	X	X	X
Canada	CEQUEAU Model (CEQUEAU)	X	X	X	X	X	X
Czechoslovakia	Empirical-Regressive Model (ERM)	X		X	X	X	X
Denmark	Rainfall-Runoff Model, Version II (NAM-II)	X	X	X	X		
Japan	Tank Model with Snow Model (TANK)	X	X	X	X	X	X
Sweden	HBV-Model (HBV)	X	X	X	X	X	X
Switzerland	Snowmelt Runoff Model (SRM)	X	X	X	X		
United Kingdom	Institute of Hydrology Distributed Model (IHDM)		X		X		
USA	Streamflow Synthesis and Reservoir Regulation (SSARR)	X	X	X	X	X	X
USA	Precipitation-Runoff Modelling System (PRMS)	X	X	X	X		
USA	National Weather Service Snow Accumulation and Ablation Model (NWSRFS)	X	X	X			

Table Xa

The graphical and numerical verification criteria used in the first project

Graphical criteria	*Numerical criteria*	*Flow variables for which numerical criteria were indicated*
1. Linear scale plots of simulated and observed hydrographs 2. Double mass plots of simulated versus observed monthly discharge volumes 3. Flow duration curves of simulated and observed daily discharges 4. Scatter diagrams of simulated versus observed monthly maximum daily discharges (peak flows)	1. Coefficient of variation of residual of errors given by: $Y = \dfrac{\left[\dfrac{\Sigma (y_c - y_o)^2}{n}\right]^{1/2}}{\bar{y}_o}$ 2. Ratio of relative error to the mean given by $R = \dfrac{\Sigma (y_c - y_o)}{n\bar{y}_o}$ 3. Ratio of absolute error to the mean given by $A = \dfrac{\Sigma \lvert y_c - y_o \rvert}{n\bar{y}_o}$ 4. Arithmetic mean given by $D = \dfrac{\Sigma y_{o,c}}{n}$ 5. Phasing coefficient (PH) for the monthly peak flows given by the number of times that the simulated peak is shifted in time from the corresponding observed peak by at least oneday 6. Coefficient of persistence (PE) given by $\dfrac{\sum_{i=1}^{k} B_i^2}{V}$	1. **Mean daily discharge (in $m^3\ s^{-1}$)** 2. **Maximum daily discharge ($m^3\ s^{-1}$) for each month in which the flow equals or exceeds the mean flow for the whole period of calibration and verification** 3. **Monthly volumes of flows expressed in centimetres of depth over the cathment area** 4. **Mean daily discharge for low flow days defined as those days during which the flow is above zero and is below the flow not exceeded during a period of 130 days during the verification period. For the Wollombi Brook data set, this variable was defined as the flow below the mean for the whole verification period.**
	In the above equations: y_o = observed discharge y_c = computed discharge n = total number of observations $\bar{y}_o = \dfrac{\Sigma y_o}{n}$ k = number of positive and negative runs $V = \sum_{i=1}^{n} (y_o - y_c)^2$ for the n items B = the individual areas of each segment	

period to produce a simulated discharge (computed output). The simulated discharges produced by the tested models for both the calibration and verification periods in each data set were then centrally evaluated and compared by WMO using several graphical and numerical verification criteria agreed upon by all modellers. On the basis of this evaluation and intercomparison, several conclusions and recommendations were made concerning the performance and use of the models in various forecasting situations.

(c) *Recommendations on verification criteria*

(i) As one of the many factors involved in model selection, it would be advantagous if the verification and intercomparison of models in general could be carried out in accordance with at least some generally accepted verification criteria. The numerical verification criteria for such general use should, as far as possible, be selected from among those used in the WMO project.

(ii) With respect to the graphical verification criteria, the double mass curves of observed versus simulated flows and in particular the flow duration curves should be available for all verification and intercomparison procedures as they convey a maximum of information.

(iii) As an alternative to double mass plots a curve of the sums of the residual errors versus time can be used. The sum of residual errors is expressed as: $(y_c - y_o)$ where y_c and y_o are the computed and observed discharges respectively. This curve, plotted on the same graphs as the observed and simulated hydrographs, assists in the analysis, particularly with respect to the model's ability to keep account of volumes.

The graphical and numerical verification criteria used in the evaluation of the simulated discharges produced by the tested models are given in Table X(a).

(d) *General conclusions and recommendations*

(i) Criteria for selection of models: The selection of a model for a specific forecasting situation should be guided by the following criteria:

- General and specific purpose and benefits of the forecast (e.g., continuous hydrograph of discharges, floods, water quality, water resources management, etc.);

- Length of record of the various types of input data;

- Quality of data field, both in time and space;

- The availability and size of computers, for both development and operation of the model, as well as the possible use of the model by relatively non-expert hydrological forecasting personnel;

- The possible need for transposing model parameters from smaller catchments to larger catchments, usually downstream where sufficient data for development are not available, and the possible application of models to larger river systems with important human interference (man-made structures);

- The ability of the model to be conveniently updated on the basis of current hydrometeorological conditions.

(ii) River catchment characteristics

- With respect to climatic and physiographic characteristics of the basin, a prospective user of a model does not need to be highly discriminating in the selection of models if the basin for which the model is selected is climatically humid and a large supply of moisture is available throughout the forecasting and development period. In climatically humid basins simple models may serve just as well as complex, sophisticated models. For basins in semi-arid regions the prospective user of the model will have to be highly discriminating in the selection of the model.

- The simulation of streamflow by use of explicit moisture accounting models is dependent upon characteristics developed to represent the water balance for the entire period of record. In general, explicit moisture accounting models are better equipped to simulate river response during and after a long dry spell. However, such models frequently demonstrate the undermeasurement of precipitation which is a feature of severe storms. Therefore, the tendency to underestimate individual large flow events (peaks) demonstrated by the explicit moisture accounting models could be alleviated in actual operational forecasting procedures by objectively modifying reported precipitation input values.

- The implicit moisture accounting models, such as the TANK, are very adaptable and flexible with respect to the size and climatic and physiographic characteristics of the basin. In the TANK model, for example, by introducing additional reservoirs (tanks), stream channel elements and areal distribution of soil moisture may be simulated if such components are predominant in the runoff process.

- In the presence of poor quality data for model development, implicit moisture accounting models, and in particular index (system approach) models, may have a better capacity to cope with this deficiency and therefore may give better forecasting results than explicit moisture accounting models.

(ii) Data sets and parameter optimization

- When several models are being compared on several data sets the results are more meaningful if all the models are tested on all the data sets. Provision should therefore be made for all data sets to satisfy the requirements of all models.

- With respect to parameter optimization, or the fitting of a model to a particular basin, two basic methods, manual and automatic, are available. Manual methods may produce a good set of parameters. However, in addition to the length of time required in terms of man hours and degree of interplay with the computer, the hydrologist performing the optimization must necessarily possess a large degree of skill and experience with the model being used and the basin on which it is applied. Automatic methods, on the other hand, are fast and simple to use. But they have some inherent disadvantages. Some of these are: a complete dependence on one objective function and sub-optimal solutions due to the concavity of portions of the response surface and poorly selected initial parameter values, and failure to recognize the effect of perturbing a group of parameters simultaneously. This may result in pure curve fitting and produce a set of parameters which fit the calibration data reasonably well, but are hydrologically unrealistic; subsequently the model may give poor results in actual forecasting. As far as possible, a combination of manual and automatic procedures should be used in model calibration.

The snowmelt-runoff models comparison

Models involving snowmelt were not included in the first project, one reason being that the data sets used for the comparison of performance were not designed for that purpose. A second project, designed specifically to compare models of snowmelt runoff, was therefore undertaken during the years 1976 through 1983.

(a) *The aims of the second project (WMO, 1986) were:*

(i) to obtain information on models used for forecasting snowmelt runoff and to arrange for an intercomparison of such models;

(ii) to report on the results of the intercomparison and thus offer guidance on the use of the models in various situations; and

(iii) to disseminate to interested countries the material prepared in order to encourage new approaches to forecasting snowmelt runoff and to assist countries in which it is planned to use models for such forecasting.

The object of the project was not to find the model which fits best in all circumstances but to give to users the opportunity to examine the performances of the tested models under various conditions.

The project involved the intercomparison of 11 operational models of snowmelt runoff submitted by eight countries. The models were tested using data sets for six river catchments from climatologically and geographically varied conditions in six countries. Each data set consisted of two distinct periods; a calibration period (six years) and a verification period (four years) following immediately thereafter. The model owners were supplied with observed input data (precipitation, evaporation and other meteorological data) and observed output data (streamflow) for the six-year calibration periods and only observed input data for the four-year verification periods. The observed output data for the four-year verification periods were retained by the WMO Secretariat.

For each data set the model owners used the concurrent observed input and output data for the six-year calibration period to establish the parameters of their models and employed the additional four years of observed input data in the verification period to produce a simulated discharge (computed output). The simulated discharges produced by the models for both the calibration and verification periods in each basin were then evaluated and compared by the WMO Secretariat using graphical and numerical verification criteria previously agreed upon by the model owners.

A technical conference was organized by WMO in Norrköping, Sweden, in September 1983. It was attended by representatives of agencies participating in the project and several invited experts. The conference considered the results of the project and prepared a number of conclusions and recommendations which are summarized herein and which are presented in full in the report of the project (WMO, 1986).

(b) *Models*

Most models of snowmelt runoff consist of two components: a snowmelt model, which simulates the process of snow accumulation and melting, and a transformation model which takes the snowmelt or the rainfall(as input data), and yields the basin runoff as output.

At the start, the model owners proposed a tentative classification of both the snowmelt and transformation parts of their models.

A preliminary review was also made of which physical processes are taken into account by each model. The final report contains a description of the various algorithms and equations used to simulate these processes, presented under the following headings:

(1) Handling of meteorological data:

- (i) Method of subdividing catchment;
- (ii) Distribution of temperature (lapse rates);
- (iii) Determination of form of precipitation;
- (iv) Distribution of precipitation (spatial variability, elevation effects and methods of correcting precipitation measurements);
- (v) Distribution of other meteorological inputs.

(2) Structure of snowmelt models:

- (i) Accumulation (interception, depth and density of snow);
- (ii) Areal snow distribution;
- (iii) Surface energy exchange (including seasonal variation);
- (iv) Internal processes (the cold content of a snowpack, liquid water storage and percolation);
- (v) Snow-soil interactions.

(3) Structure of transformation models.

Information has also been assembled and published on the data requirements of the models and on the methods used to fit them during calibration.

(c) *Data sets*

Table IXb indicates which data sets were used by each model.

The model owners were requested to send to the WMO Secretariat the output computed by their models for both the calibration and verification period of the data sets on which they had run their models. The computed output was presented for both the streamflow runoff at the outlet of the tested catchment and the total snowmelt over the catchment. If, from the computational procedure, the computed snowmelt normally included rainfall, figures for snowmelt and rainfall were presented separately where possible.

(d) *Verification*

Three graphical and nine numerical verification criteria were used in this project to evaluate and compare the simulated discharges:

Graphical verification criteria

- Linear plots of simulated and observed daily discharge for each year of the calibration and verification periods;

- Flow duration curves (i.e. runoff/average runoff) as a function of time) for accumulated and observed daily discharges. Separate curves were plotted for the calibration and verification periods;

- Scatter diagrams of simulated vs observed monthly maximum daily discharge (peak flows), separately for the calibration and verification periods.

It was considered that the linear scale plot of simulated and observed daily discharge was the most important graphical verification criterion.

Numerical verification criteria

The numerical verification criteria used in the project are given in Table Xb.

Verification results

Several examples demonstrated the variability of the numerical values of the criteria of fit. It is important to realize the limited description of overall performance that can be contained in one number. In order to obtain some information about the variability of the criteria, four such criteria (NTD, S, R and NS) were also computed for each individual year and individual snowmelt season in the calibration and verification periods.

(d) *Intermediate outputs*

An upper limit of 5000 km^2 was fixed on the size of the river basins used in the project as one means of reducing the effect of the transformation model on the overall response.

Table Xb

The graphical and numerical verification criteria used in the second project

- Ratio of standard deviations of computed to observed discharges:

$$CO = \sqrt{\frac{\sum(y_c - \bar{y}_c)^2}{\sum(y_o - \bar{y}_o)^2}}$$

- One minus the ratio of the sum of squares of the daily residuals to the centered sum of squares of the daily observed discharges:

$$NTD = 1 - \frac{\sum(y_c - y_o)^2}{\sum(y_o - \bar{y}_o)^2} \quad \text{for mean daily discharges}$$

- Ratio of the quantity (1-NTD) for the verification and calibration periods:

$$R1 = \frac{(1 - NTD)\ VER}{(1 - NTD)\ CAL}$$

- Ratio of the sum of squares of the monthly residuals to the centered sum of squares of the monthly observed discharges:

$$NTM = \frac{\sum(y_o - \bar{y}_o)^2 - \sum(y_c - y_o)^2}{\sum(y_o - \bar{y}_o)^2}$$

- Ratio of the standard deviation of the residuals to the mean observed discharges:

$$S = \frac{\sqrt{[\sum y_c - y_o]^2/n}}{\bar{y}_o}$$

- Ratio of the mean error to the mean observed discharge:

$$R = \frac{\sum(y_c - y_o)}{n\ \bar{y}_o}$$

- Ratio of absolute error to the mean observed discharge:

$$A = \frac{\sum|y_c - y_o|}{n\ \bar{y}_o}$$

- Coefficient of gain from daily means:

$$NS = \frac{\sum(y_o - \bar{y}_{od})^2 - \sum(y_c - y_o)^2}{\sum(y_o - \bar{y}_{od})^2}$$

- Ratio of sum of absolute errors to the total observed runoff volume:

$$PD = \frac{\sum_{i=1}^{N} \left| V_{oi} - V_{ci} \right|}{\sum_{i=1}^{N} V_{oi}}$$

As stated above, the model owners were invited to submit computed output for total snowmelt in the basin. All but one of the model owners submitted such intermediate output and these outputs were seen as being of great potential interest to those who would wish to study the detailed results of the project. The intermediate outputs computed by the tested models have therefore been included in the final set of data assembled as an outcome of this project.

(e) *Conclusions and recommendations*

Several recommendations concerning the development of models, optimization techniques, and the requirements for verification criteria and the preparation and transfer of data sets were made for use in similar projects in the future.

- There was considerable variety in the structure of the models and in the manner in which they were calibrated, as well as in their updating ability and in the degree to which they had been or could be used for real-time forecasting. It was clear that the major differences in the models derived from the different purposes for which they had been developed and the data and computing resources which were available for their use.

- The temperature index approach was applied by all of the tested models which were used for operational forecasting. This was because the temperature index approach generally produced good results and because there was a lack of real-time data for the alternative energy budget approach.

- In most models, an additional sub-routine was needed to suppress runoff during the snow-ripening period. Because of their interactions, the internal processes of heat flow (charge in cold content), liquid water storage and percolation of free water could be modelled collectively as a group.

- The six data sets used in the project were considered to be generally satisfactory. They satisfied the needs of most of the models tested in the project.

- If at all possible, data from more than one precipitation and temperature station should be acquired for a river basin Not only do two or more stations provide a measure of cross-checking, they also give areal information required by most meteorological analyses. The same considerations hold true for other variables such as wind speed and snow data;

- The subdivision of river basins into elevation zones was considered desirable because of the strong elevation dependent gradients of temperature and precipitation in mountainous areas;

- A number of models used precipitation adjustment factors to account for systematic measurement error, lack of areal representativeness of the point measurements, and interception and sublimation losses. When possible, correction should be made for systematic measurement error before making further adjustments;

- It was important that a range of verification criteria be evaluated in order to cover different aspects of model performance. However, not all the criteria proposed for use in the project had supplied information that could be usefully interpreted in terms of model performance. The three graphical criteria and the nine numerical criteria finally adopted for the project might be used as a basic set in future projects of this nature.

- WMO should consider the possibility of conducting a simulated real-time intercomparison of rainfall-runoff and snowmelt-runoff models designed to compare their performance under streamflow forecasting conditions with updating allowed

Guidance to a HFS designer

The conclusions and recommendations of the two model comparison projects provide a guidance to designer of HFS for the selection of models. It should be stressed, and the author's experience strongly confirms it, that in each HFS several models should be tested in real operational conditions. These should be selected among the simplest and more complex ones. Two or three models are appropriate, in addition to a "manual" back-up forecasting procedure to be applied in case of failure of the computerized equipment. Next section is eqully important for the selection of models, as it demonstrates the relative dependence of the different components of the HFS, particularly showing that the importance of the model should never be exaggerated.

The experience of the HFS installed in Burma provides yet another confirmation of the above views of the author. The Burma Meteorological and Hydrological Service installed four models for forecasting on the country's large rivers, but in particular on the Chindwin in the Irrawady basin complex. The models are: Sacramento, COSSAR (a simplified version of SSARR), TANK of Prof. Sugawara and the API (Continuous), a model originated in USA and used by WMO in several projects. While the resulting forecasted flood hydrographs are somewhat different, the relative errors ascertained by using numerical criteria of the WMO project on rainfall-runoff models comparison, are of a magnitude, which is insignificant if compared to errors in measurement of both rainfall and discharges, particularly during floods.

More detailed information on forecasting models

All the types of models classified and described above are presented in detail in the references. There are at least two recent publications including a detailed description of forecasting models.

The first "River Flow Modelling and Forecasting" edited by D.A. Kraijenhoff and J.R. Moll (1985) contains a number of contributions from a number of European and U.S. authors on available forecasting techniques and a few contributions, including one of this author, on the design of forecasting systems in real world. The second one, "Hydrological Forecasting" edited by M.G. Anderson and T.P. Burt (1985) contains contributions from the United Kingdom and USA on different types of models and modelling problems not only of water quantity but also quality including groundwater. With the exception of a contribution on principle of radar measurement of precipitation, the real world "forecasting systems" have not been included in the six hundred pages of this publication.

The experience of this author indicates that no model has ever been used in a HFS on the basis only of information provided in scientific papers or compendia similar to the two mentioned above. This has influenced also his approach to the description of available models in this part of the text. As long as the model is intalled on a computer, in addition to the complete computer program of the model, the data base management software has to be available for the real world HFS. The above publications seem to consider this requirement as a detail, superfluous even to be mentioned.

To the knowledge of this author the only readily available means of transfer of real-world forecasting models is the WMO HOMS system, which does provide also for the most often indispensable consultant assistance in the transfer and training in the use of the modelling and data base management software.

3.2 SENSITIVITY AND INTERDEPENDENCE OF HFS SUB-SYSTEMS

The correct design of a HFS requires an evaluation of the significance of all the sub-systems (components) and should not be biased toward one of them, which is very often the case. Indeed, many hydrologists and as a result managers of HFS consider the forecasting procedure or "the model" as the most important component and often pay little attention to the interrelation of this component with others, in particular the real-time data collection sub-system.

The two sub-systems mentioned above are often considered separately, at least as far as their influence on the final result of the forecast is concerned, although undoubtedly everybody agrees that they have to be technically compatible and properly interfaced. What is often less evident is that the data collection sub-system may be crucial for the accuracy and cost-effectiveness of the HFS and that its design or selection is at least as important as that of the hydrological model.

The above-mentioned survey in Europe has clearly indicated that the majority of the data inputs used in HFS are those on precipitation and discharges, with the understanding that, in systems using only discharges as input, the above statement on relative significance of the components of HFS applies only to discharge data collection. The stream gauging, however, is fortunately, by and large, the most accurate of all measurements carried out at basin level, as it integrates the runoff over the entire area and does not involve area sampling. According to Herschy (1978), if carried to relevant ISO/WMO standards, the upper limit of error at 95% confidence is for single determination of discharge $\pm$ 7%, for standard error of the mean of the stage discharge relation is $\pm$ 5%. Similar limits of errors are for daily, monthly and annual discharges. It is obvious that in difficult conditions of developing countries and in floods these errors may be significantly larger; 15 % errors at 95% confidence for flood peaks are not uncommon but they are still minor in comparison with errors in precipitation and evaporation measurement and their areal estimations. On the other hand, the relative significance of precipitation and evapo-transpiration data for discharge computation by conceptual models (the Sacramento model) has been ascertained by Nemec and Schaake (1982) and illustrated by Figs. 14 and 15, both in a dry and a wet basin.

From the above, it appears that the single most important input data in HFS using a rainfall-runoff model is the value of the rainfall. This is clearly demonstrated also by a study of sensitivity to inputs by conceptual hydrological models, in particular of the Sacramento model. Figure 16, based on a study by Zettlemoyer and quoted by Burnash et al (1981), indicates that a five per cent change in the value of the input rainfall produces an

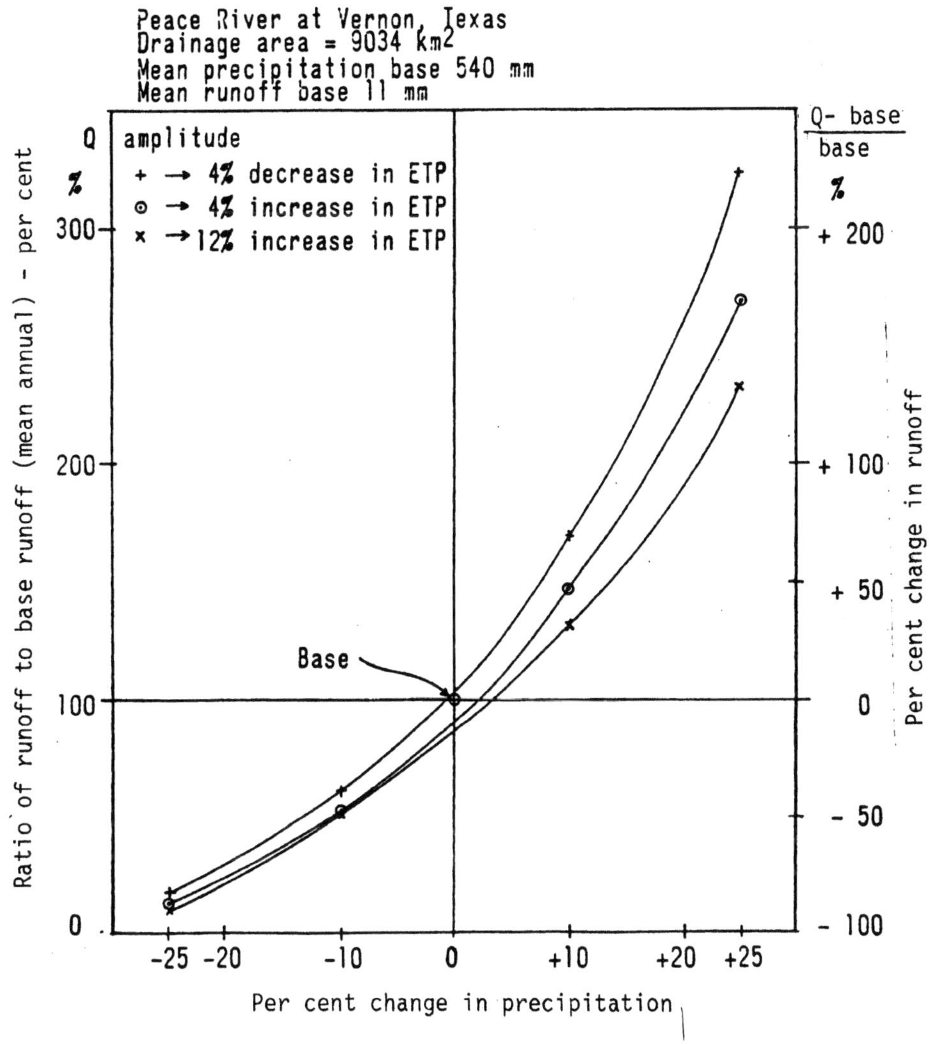

Figure 14. Change of runoff resulting from change of precipitation and evapotranspiration (arid climate conditions).

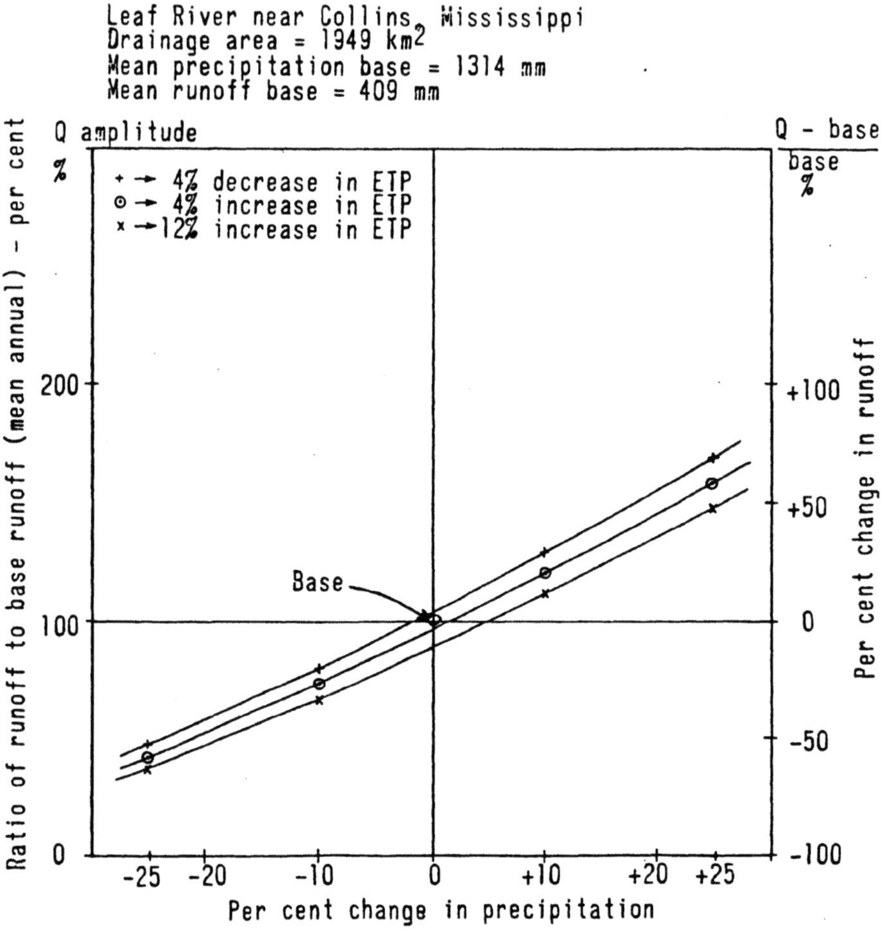

Figure 15. Change of runoff resulting from change of precipitation and evapotranspiration (humid climate conditions).

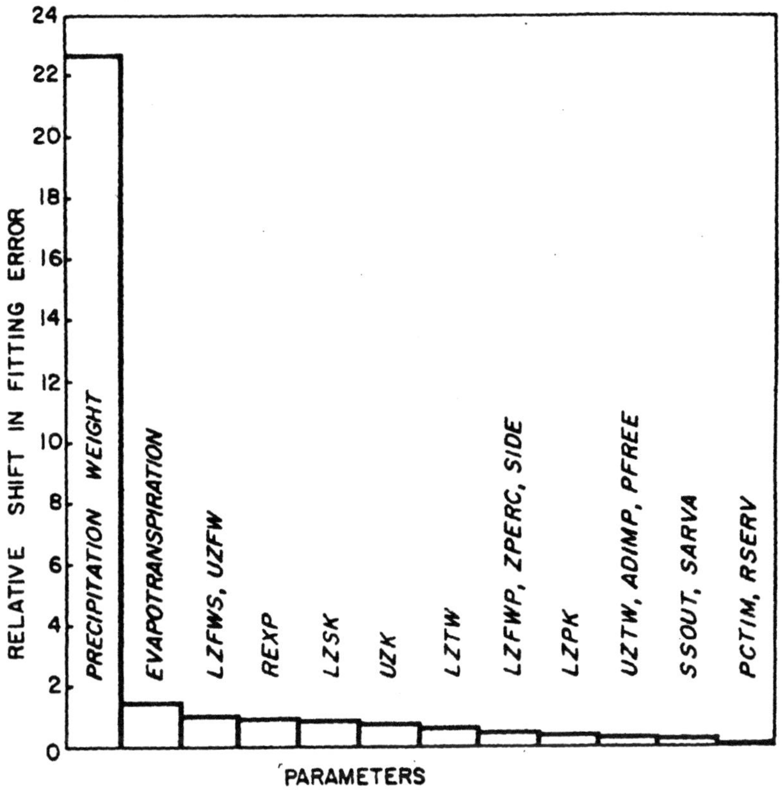

Figure 16. Relative change in streamflow resulting from a shift of modelling parameters.

impact on the forecast of runoff nearly an order of magnitude higher than a similar shift in any other parameter or input, including evapo-transpiration.

Unfortunately, it is not easy, if at all possible, to ascertain the "real" value of rainfall over the basin. Many studies of this problem have been conducted all over the world (see WMO 1973). This value is subject to error, not only related to the instruments but also to the method of evaluation of the areal sampling, see for example the work of Burnash et al (1981) quoted above. For the analysis of the relative significance of the HFS sub-systems, it is, however, possible to assume that the value adopted and corroborated by statistically optimal simulation of discharge, as tested against that observed, is the "true value". The sensitivity of the results to the two sub-systems could then be expressed by the change of the discharge in passing from one model to another with the same simulated "error" of the input rainfall.

Such tests were undertaken with two models used in one basin in Africa. The first model was the Sacramento model (Burnash, Ferral, McGuire, 1973), the second was the so-called "rational formula" which translates rainfall into runoff by a single runoff coefficient, which is established as an average ratio of rainfall and runoff during the whole period of observation. A simple unit hydrograph is then used to distribute the runoff in time.

It is obvious that, in a natural catchment, this last "model" grossly distorts the flood events by underestimating discharges in wet conditions and overestimating them in dry conditions.

The basin used for the test was in the larger basin of Lake Victoria in Africa, where data were very unreliable and covered only seven years of observations. A total of 13 flood events were used for the tests. The results are in Fig. 17. It is clear that the less sensitive "model", the rational formula, will also be relatively less influenced by the "error" in the precipitation, as can be seen from Fig. 18. It is also seen from this same figure that the crude model results in larger absolute error, with the "true" rainfall observation. It is, however, very clear that the "error" in rainfall observation may entirely cover the relatively large increase in accuracy resulting from the use of a good conceptual model. Indeed, if the rainfall average over the catchment is underestimated by 25%, which is not unusual, both the Sacramento and Rational Formula give, on the average, the same value of peak discharges. On the other hand, with "true" value of precipitation, the relative accuracy of the Sacramento model is better by a factor of three. Burnash et al. (1983) conducted a similar test on a much better observed and well-instrumented basin of the Eel River in California, for which observations for 20 years permitted a fitting of the Sacramento model resulting in a correlation between simulated and observed flow of .99. The

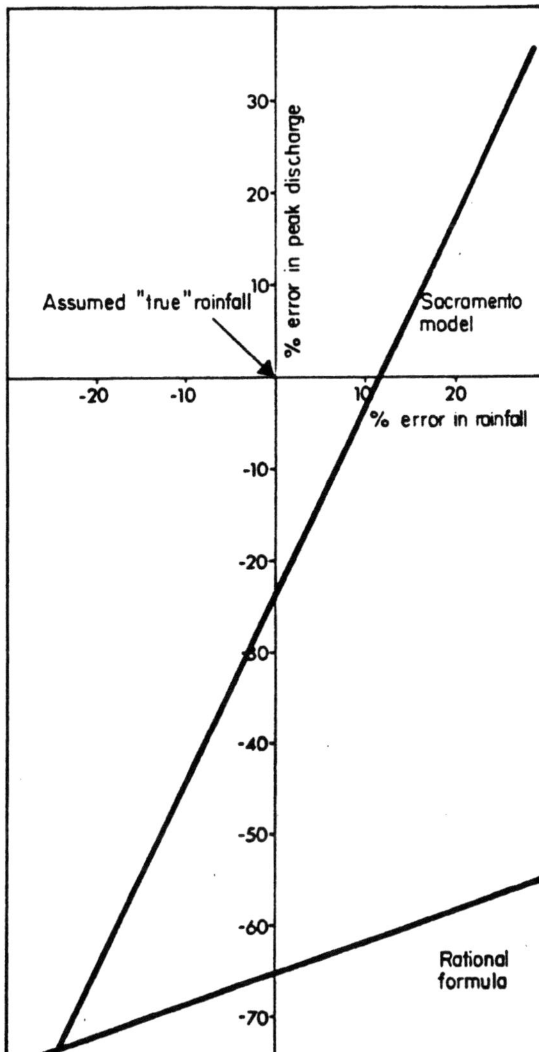

Figure 17. Effect of rainfall "errors" on predicting peak discharge (Lake Victoria catchment, Uganda, Africa).

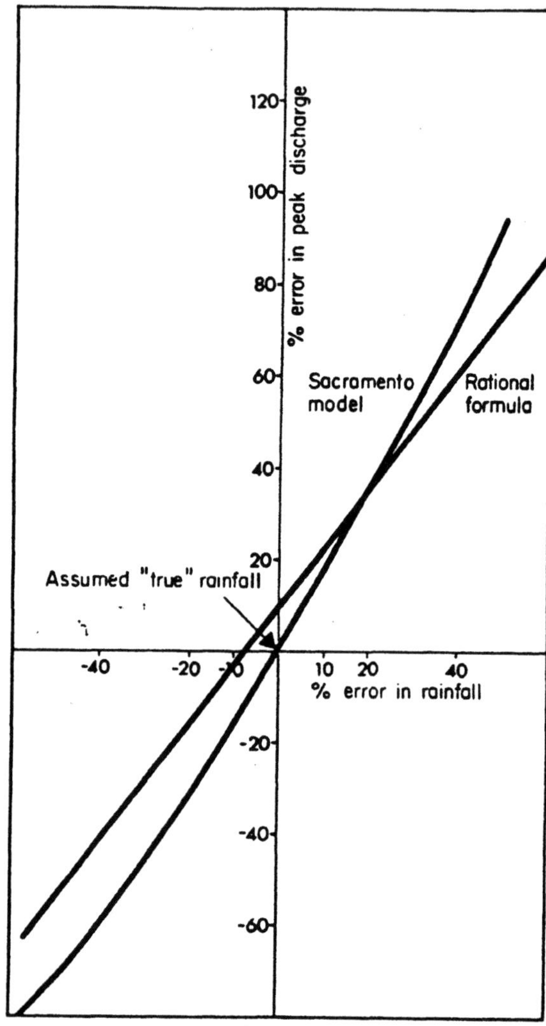

Figure 18. Effect of rainfall "errors" on predicting peak discharge (Eel River, California, USA).

well-calibrated conceptual model gives excellent results with the "true" precipitation (see Fig. 14). It is again more sensitive to the simulated "error" rainfall values over the catchment than the very elementary "rational formula". However, should the rainfall "error" be an increase of 20% over the "true" value, both forecasting techniques give the same results and are in error by about 35% above the correct observed value for the peaks. The conclusions of the above tests indicate that a concentration of effort to improve the forecasting technique has significance only if the rainfall observing system is sufficiently reliable and the data available for the fitting of the model are such as to avoid a large value of absolute error in all models. In other words, a designer of a HFS has to keep in perspective all components of his system design and not concentrate attention exclusively on his model.

Another question closely related to the above is whether a significant cost-benefit or cost-efficiency improvement of HFS can be achieved by adding to a rainfall-gauge network a rainfall-measuring radar.

The conditions in which the use of radar in a HFS have been discussed in detail in section 2.2. It is recalled that this use, while most important in areas of tropical cyclones, does not look favourably from the cost-efficiency perspective in most of the developing countries. It is not recommended by this author in cases where the radar's purchase is motivated exclusively by purposes of a HFS. In the experience of the author, particularly in developing countries, the cost-efficiency factor would weigh heavily against the radar-cum-gauges in favour of the telemetering raingauge system only. This conclusion has been recently corroborated both by Ishizaki (1979) and Sugawara (1981).

A sensitivity analysis can be made for other components (sub-systems) of a HFS. Nevertheless, the data collection on one side and the hydrological model on the other are, in the opinion of this author, the most sensitive sub-systems with respect to their relative interdependence. They should never be considered separately and a decision on their selection should not be taken without at least a summary cost/benefit or cost/effectiveness analysis.

Similar analysis of interdependence of HFS sub-systems, in the opinion of this author, may provide the design and management of a HFS as an entity without technically unwarranted bias. However,the author recognizes that a large number of non-technical considerations may influence the design and are often unavoidable. Thus, as in many similar situations, a technician provides his evaluation and then hopes for the best.

The experiences gained in a large number of HFS installed in South-East Asian area prone to tropical cyclones (typhoons) within the framework of the Typhoon Operational Experiment (TOPEX) demonstrate very clearly the need of a balanced approach to the design of HFS (in the case of the TOPEX the HFS were exclusively designed for flood forecasting and can be thus abbreviated as Flood Forecasting Systems - FFS).

TOPEX involved three distinct sets of activities, undertaken under:

(a) the Meteorological Component;
(b) the Hydrological Component; and
(c) the Warning Dissemination and Information Exchange Component.

The Meteorological Component concentrated on a Core Experiment which was designed to obtain data on typhoons from an intensified observational network and to upgraded routine operational work in the exchange of such data and on the analysis and forecasting of typhoons. A sub-experiment was implemented to obtain detailed information for use in research on typhoons. The Hydrological Component concentrated on the assessment in real-time of the performance of existing FFSs, the study of various aspects of these systems with a view to improving their performance and aiding the design of new systems, and the further development and application of techniques for flood risk evaluation. The Warning Dissemination and Information Exchange Component was concerned with the dissemination of warnings required for the safety of life and property and the exchange of related operational information. Activities under the hydrological component may be summarized as follows:

(a) The evaluation of established flood forecasting systems through the monitoring of their performance during flood events and the study of the accuracy of the forecasts made;

(b) The identification of the forecasting models used and a comparison of their performance in real-time forecasting operations;

(c) The evaluation of the separate and/or combined hydrological effects of typhoons, particularly river and storm surge flooding, and thus the determination of associated flood risk.

Countries were invited to designate river basins (DRBs) whose FFSs would be evaluated. A total of nine river basins were designated by six countries, as follows:

- Puyang (China)
- Xizhi (China
- Yodo (Japan)

- Kitakami (Japan)
- Kelantan (Malaysia)
- Pampanga (Philippines)
- Bicol (Philippines)
- Han (Republic of Korea)
- Pasak (Thailand).

Insofar as the inhabitants and the economy of any particular river basin are concerned, the source of the precipitation which causes flood damage is almost irrelevant, whether it be typhoon, monsoon rainfall or snowmelt. Any FFS must be designed to cope with all the types of event which may cause flooding in that particular basin. All the DRBs in the TOPEX study can suffer significant floods from causes other than typhoons, and in some cases (e.g. Malaysia) the worst floods appear to occur outside the typhoon season. None of the countries appear to have developed their HFS to deal solely with typhoon events. It would, therefore, be illogical to judge any or all of the HFS on the basis of typhoon events only. Accordingly, the evaluation of the various FFS was made for all types of precipitation which can cause flooding in the basins under consideration.

The models installed in the different FFS were as follows:

Country	Basin	Soil Moisture* accounting	Hydrograph of flood	Routing
China, P.R.	Zhui (Puyang)	IZ	UH	MSKG
	Yizhi	IZ	UH	MSKG
Japan	Yodo	SF		SF
	Kitakami	SF		SF, StV
Malaysia	Kelantan	TANK	SACRAM	SACRAM
Philippines	Pampanga	TANK	UH, SF	SF
	Bicol	TANK	UH,SF	SF
Thailand	Pasak			MSKG
Korea	Han	SF		SF

*The abbreviations of models used for the three different functions of a rainfall-runoff model (soil moisture accounting, distribution of runoff in a hydrograph, routing through channel network) are as follows:

IZ	=	A simple one layer model of soil
UH	=	Unit Hydrograph
MSKG	=	Muskingum (hydrologic routing)
SF	=	Storage function (hydrologic routing)
StV	=	Dynamic wave by St. Venant equation (hydraulic routing)
TANK	=	Tank Model of Sugawara
SACRM	=	Sacramento Model

The TOPEX hydrological forecasting experiment was, to the knowledge of the author, the most extensive operational testing of HFS carried out until then. It consisted in:

I. Evaluation of designated FFS with consideration of the following factors:

1. Collection of field data

(a) Density and location of observation stations
(b) Frequency of observations
(c) Reliability of observations

2. Transmission of field data to forecasting centres, including reliability and timeliness of transmission

3. Processing and analysis of data

4. Accuracy of forecast

5. Dissemination of forecast, including its timeliness

6. Use made of meteorological forecasts, including the utility of inputs from the meteorological component of TOPEX.

II. Identification and comparison of forecasting models

1. Number and types of models identified

2. Number and types of models compared and tested

3. Comparison of models

(a) Applicability of models
(b) Accuracy of results
(c) Timeliness

III. Flood risk analysis

1. Flood risk analyses and studies carried out and flood risk maps prepared in the past and during the TOPEX period by each country

2. Actual use of above studies and maps in operational flood forecasting and in determining the risk of flooding from rivers

3. Use of inputs from meteorological component of TOPEX in these studies

IV. Storm-surge flooding analysis and forecasting

1. Number of studies on storm-surge flooding in the past and during TOPEX period carried out by each country

2. Use of inputs from meteorological component of TOPEX in these studies

It has been said that the correct design of a FFS requires an evaluation of the significance of all the sub-systems (components) and should not be biased towards any one of them, which is very often the case. Effective operation of the FFS, involving human communication and organizational factors, is not by any means guaranteed, even if it correctly designed, as the potential exists for deficiencies in human communication and organization.

The principle criterion for evaluation of the individual FFS in TOPEX is therefore the "balance" between all the various factors, in relation to the scale of development in each DRB, rather than the technical or scientific expertise of only one or two components.

The evaluation of the experience gained in TOPEX was summarized in the following conclusions and recommendations:

(a) Even a brief consideration of the topographic and hydrological characteristics of a basin, using to the full recorded data where available, can provide valuable preliminary guidance on the type(s) of forecasting system that could be most effectively applied.

(b) The definition of warning levels, as triggers for issuing forecasts to the recipients, is the major criterion to be considered, and will fundamentally influence the design of a cost-effective FFS, by defining carefully the lead-time needed to ensure effective preventive measures or evacuation of people from the areas at risk.

(c) For any FFS, the necessary lead-time should be achieved by the most economic methods at which point any further investment in increasingly complex FFS achieves no real benefits.

(d) In many cases, regular observations from a reliable rapid telemetry systems coupled with stage/stage correlations will achieve the necessary objective and lead-times without the use of more complex models.

(e) Where models are used, the choice is not usually a critical factor. During the saturated ground conditions which apply in large floods, the TOPEX results indicate that simpler models can, in many cases, perform as well, if not better, than the more complex models.

(f) The quality of input data is usually far more important, and this partly depends upon proper planned maintenance of telemetry equipment. If properly maintained, this equipment can last well over ten years, provided adequate spares are available, but it can last for as little as one or two years if not maintained.

(g) The highly automated and sophisticated systems can yield good results, with the ability to develop multiple forecasts very rapidly. However, the apparent precision offered by sophisticated equipment and complex models should not be allowed to lead to a false sense of confidence in the results. An experienced forecaster is needed to interpret the results and prepare the final forecast. Considerable difficulty is often encountered in retaining personnel for long enough to acquire and apply such experience.

(h) It must not be assumed that the transfer of technology from a technically advanced information-oriented country to a developing country can proceed without problems. Where labour costs are low, it is often efficient to use a large number of outposted personnel for data collection, validation and transmission.

(i) Every sophisticated FFS should have a simple back-up system which enables some forecasts to be made if the main system fails or malfunctions.

(j) Only certain types and a limited amount of information can be understood and effectively used by the general public or by the public authorities. To issue more information than can be used may create confusion and be counter-productive.

(k) While the terminology and number of warning levels used differed between systems, the same general approach to the issuing of warnings had been adopted by all participants in TOPEX. The specific details for each system depended on the socio-economic characteristics of the river and its flood channel, especially the form of cross-section (e.g. V-shaped valley, deep channel/flat flood plain, extensive system of high level banks, etc.). The specific details of the warnings might even vary between different reaches of the same river.

(l) Some circumstances call only for forecasts to be issued indicating when the river is expected to reach a certain critical level or levels. Others require that forecasts be issued at regular and frequent intervals indicating how the water level will vary and at what rate in the coming hours or days.

(m) Much useful information can be obtained by conducting detailed surveys of the river channel and flooded areas immediately after the flood has receded. This can be particularly useful for assessing the adequacy of the warning levels used.

(n) The comparative studies of different models yielded interested results which substantiated the view that there is no one best model. Different models work best under different circumstances and in different basins.

(o) Flood risk maps are very valuable, if not essential, for use in conjunction with FFSs. Regular flood exercises have a valuable function in ensuring that all participants understand how to act and react when the extreme event occurs.

(p) There is no single measure that can be used to describe the performance of a FFS. It is the system's, and not just the model's performance, that is to be described; account should be taken of what forecasts were issued, with what lead time and accuracy, whether they proved to be necessary and whether they lead to the saving of life and reduction of property damage.

(q) The development of links with Meteorological Services through the provision of meteorological information (including QPFs) to the flood forecasting centres may improve if better specification of requirements for particular circumstances could be provided. For example, there are many basins where the lead-times based on observed rainfall or river levels/flows are more than adequate for effective operation of FFSs, and QPFs would give no real benefit. In other very specific areas, QPFs are urgently needed as the only means of achieving adequate lead-times. Meteorological research activities should be directed specifically towards these basins in order to obtain maximum benefit.

(r) The links that exist with the Meteorological Services or between hydrologists and meteorologists in the same service have proved most valuable during the three years of TOPEX. The meteorological information thus provided is always of value in giving advanced warning of a potential flood threat and was often used in a quasi-quantitative manner in preparing forecasts.

(s) Due recognition should be given to the value of the links that had been established between Hydrological and Meteorological Services and between hydrologists and meteorologists, while at the same time calling for more and improved QPFs.

The above conclusions and recommendations are a summary of the experience of the designers and operators of HFS used for flood forecasting in the Far East. To the knowledge of the author, they can however be applied to all HFS in the climatic zones similar to the one of the Typhoon Committee region. Some of the conclusions, particularly those on the interdependence of the different components of a HFS and on models, are valid for all HFS, without distinction of climatic conditions.

4. Forecast updating and evaluation

4.1 FORMS IN WHICH FORECASTS ARE ISSUED

The forms in which forecasts are issued are basically set from the beginning of an HFS design by the real or potential user's need. They depend on all elements by which the forecasts are classified (see section 1.4) namely on forecast variables, type of forecast, forecast lead time as well as means and frequency of forecast dissemination.

There are two mutually interdependent elements that set the usefulness of a forecast, unfortunately their interdependence (or rather interference) is rather negative. These are the lead time and the probable error of the forecast. By the very substance of the forecast its usefulness increases with the lead time. This rule has its exceptions in some particular cases such as flood warnings. Often the most advantageous lead time for flood warnings, due to psychological and societal factors, is not necessarily the longest one, since it is the imminence of danger that makes people react to it. A rather long lead-time may cause a certain indifference in people - they get used to live with a danger that is not imminent, and take the necessary precautions later or not at all, thus increasing the damage and losses.

An increase in lead time however increases the inherent random errors in input data and thus the error of the forecasts, as is described below. As long as the expected (probable-normal or mean-random) error of the forecast is included in the form of the forecast, the relationship between the lead time and error is evident to the user. Unfortunately, particularly for flood warnings the indication of the error is not always possible or not understood by the public.

In general, forecasts or warnings can be presented in the following forms (WMO, 1983):

(a) A single expected value of the forecasted variable (e.g. crest stage, peak flow, freezing date, etc.);

(b) A forecast variable as a function of time (e.g. the flow or stage hydrograph);

(c) A probability distribution of a forecast variable expressing the spectrum of possible future values in terms of their probability of exceedence, thus accounting for uncertain future conditions;

(d) A statement of the range of values of a forecast variable in relation to the mean or normal value, e.g. "above average" or "well below that normal for the season";

(e) A warning--a message intended to warn those concerned with the expected occurrence of dangerous hydrological events such as floods, especially flash floods, storm surge, ice jam, or freezing-up of the lake or the river reach. A warning can be expressed in numerical or in qualitative terms.

These are not standardized forms of forecast issuance. A most general, although not frequent form, is a daily or weekly hydrological bulletin, useful as a source of information for a large group of users interested in the conditions over the whole basin or even larger areas. This bulletin usually includes:

(a) A hydro-synoptic chart of the basin showing, in figures, the stages, discharges, and water temperatures for rivers, lakes, and reservoirs, including the actual volume of water stored in the reservoirs. This data would refer to a given time period, using various symbols for ice conditions, the rise and fall of river stages by designated zones, and other pertinent hydrological characteristics. Alternatively, a table could be used containing all the above data related to the observation stations;

(b) A short survey of the state of water bodies, and of tendencies toward changes in their régime in the near future;

(c) Medium- and long-term hydrological forecasts.

Such a bulletin is provided mostly under routine conditions, where the forecast is used either for water management or other economic activities of routine nature. Two examples of such bulletins are included in Annexes 9 and 10 respectively. Annex 9 contains the hydrological bulletin of the Polish Meteorological and Water Management Institute for the whole of Poland, Annex 10 a hydrological bulletin issued on the current state of the flow and forecast of the low flows using basin depletion curve of the River Niger in Niamey.

A substantially different forms are used to issue emergency warnings, which are transmitted on a 24-hour or even shorter time step. The hydrologist however in these cases normally does not issue the forecast and warning to the general public but either to a specific user (for example power company operating dams) or to a governmental agency in charge of emergencies such as civil defense organization or police, special flood emergency commissions, etc. The detailed discussion of the work of such organizations is beyond the scope of this text, it is contained in publications of national or international organizations concerned with disaster mitigation and prevention. Among the international ones are the Office of the UN Disaster Relief Co-ordinator(UNDRO) and the League of Red Cross and Red Crescent Societies. A survey of activities in flood disaster mitigation and prevention is given in UNDRO (1976).

Nevertheless, useful example of civil defense activity planning in flood mitigation is given in Annex 11 (see also section 6.4).

4.2 FORECAST UPDATING TECHNIQUES AND ERROR ANALYSIS

As indicated above, every forecast is subject to an error. A forecast error, by definition, is the difference between the value of the variable as forecasted and the value observed. Both of these values are subject of error, the first due to errors in measurement (accuracy) of input data and errors (accuracy) of the forecast procedure (model), the second due to error (accuracy) of data measurement only. Thus it can be written

$e(i) = y_f(i) - y_o(i)$ where
$e(i)$ = error of forecast in time (i)
$y_f(i)$ = value of the forecasted variable in time (i)
$y_o(i)$ = value of the observed variable in time (i)

The analysis of these errors is depending on their type or characteristic (their statistical classification) and on their use. This can either be in real-time, to improve the forecast, or in hind-sight (post mortem). The first use is called updating of the forecast, the second one is generally called evaluation or verification of the forecasts.

The statistical classification of errors can be as follows (Miquel, 1984):

- "normal" errors (not to be confounded with a "normal-average" or "standard deviation") of an amplitude corresponding to the calibration of the model and of input data.
- "accidental" errors, of a large amplitude but occuring only sparsely - for example during one single flood forecast
- "persistent" errors, occuring in all forecasts (systematic underrating of peaks, for example)

A second type of classification can separate

- "independent" random errors, either fully independent or quasi-independent and difficult to be explained, and
- "correlated" errors, occuring at least in a few successive forecasts (or forecasting time steps).

These classifications can be summarized in a tabular form:

TABLE XI

Types of errors

Type of error	Origin of error: Measurement of data	Noise in input data	Model accuracy
Normal random error	-accuracy of measurement (of levels, etc.) -discharge gauging and computation error	-missing input data -reconstituted input data -errors in routine primary processing of data	-residual error after satisfactory calibration
Exceptional random error	-malfunction of an instrument -weather influence on measurement	-misplacing of data in processing -coding error	-inadequacy of used model-missing parameter (ground water, etc.)
Systematic error	-errors in rating curve (for discharges) -systematic error of instruments	-change of flow regime (for discharges) -non-stationarity of time series -seasonal influence -influence of structures in the river	-insufficient calibration independent of model adequacy

The updating and evaluation of forecasts is based on error analysis. This analysis is becoming even more important in HFS which are automated or on-line.

The activities in error analysis not only improve the accuracy of forecast but also help in the maintenance of data collection networks in establishing error detection procedures and routines and in case of failures indicates the need of switching from normal forecasting procedures to those emergency.

Depending on the character of errors, the error analysis aims at:

- reduce to a minimum "normal" random error
- identify "exceptional" random error
- ascertain systematic errors and eliminate them as appropriate.

The ultimate result of error analysis should be an improvement of forecasts but it can also have negative results depending on the adequacy of the model and initial quality of the input data and measurement of verification data (Miquel, 1984).

If the quality of data is good (the white noise is minimal) the influence of the error analysis on the forecast will be negligible. If it is very bad, the statistical hypothesis of the error analysis may become questionable and the use of the latter as well, if not detrimental to the forecast accuracy. A similar reasoning applies for the adequacy of the model, although as mentioned elsewhere in this text (section 3.2) the author does not overestimate the importance of the model for the result of the forecast. According to Miquel (1984) a maximal efficacy of updating is for "normal" but correlated, systematic errors which are not exceptionally large (do not occur suddenly and do not last only a short period). In other words updating should not be used to correct serious inadequacies of the model or occurrence of phenomena, for which the model was not calibrated. The updating in this case will not eliminate the cause of the error, indeed, in the long run, may increase it.

The simplest method of updating is an adjustment by subtraction (in relative terms) of the ascertained error e(i) from the new forecast time step values y_f (i+1), y_f (i+2), etc. To avoid discontinuities, the adjustment is usually blended into the computed variable (hydrograph) over several time steps.

The use of computers permits a much more efficient process of real-time error analysis by time series analysis methods. These ascertain correlation of errors (trends) and periodicities, which can be subsequently extrapolated into the future time steps as forecasted error e(i+1) ... e(1+n) to be added to (subtracted from) the corresponding variables y_f(i+1)...y(i+n).

The simplest identification of a systematic error (or the persistency of error) is to compute the coefficient of autocorrelation r_j of sequential forecasts. If this coefficient is not near or equal to zero, it is useful to establish a correlogram, which is a function of r_j. If the average of the errors is equal to zero, then $r_j = k^j$. It is easy to see that the simple subtraction (or addition) of the error e(i) to the next time step forecast y(i+1) assumes that k=1. If k is not equal or near to 1 the structure of the error function can be assumed, for extrapolation purposes

$$e(i) = k.e(i-1) + \sum(i) \text{ where}$$

$$\sum(i) = \text{white Gaussian noise.}$$

If the correlogram is not a simple function k^j, several methods of time series analysis can be used. The most frequent ones are the autoregressive structures of first or second order (Markov chains).

The error analysis discussed until now did not consider a separation of the errors of the input data from the errors resulting from the model. The assessment of the error function of the model may permit corrections (of the forecast) by a re-computation of its parameters. Such updating techniques are called "recursive parameter estimation" and they can be used both for the model error functions and the input data, providing that the model is relatively simple. This is the case for example of linear model such as unit hydrograph or linear systems approaches (CLS model, linear cascade of reservoirs based on hydrologic routing, etc.). The most popular "recursive parameter estimation" technique among modellers, although not so much among practicians forecasters, is the so-called Kalman filtering, applied to linear models, or extended Kalman filtering for non-linear models. This approach adjusts the input data and consequently the state variables of the model in such a way as to reproduce more closely the current and previous values of the time variables and thus updates the forecast by the change of the state variable. Extensive treatment of Kalman filtering may be found in Wood and O'Connell (1985).

Once again however it is important to recall, that the choice of the updating procedure depends on consideration of not only the type of model used, which is often presented as the most important factor, but on the amount and quality of available past data and the reliability of the real-time data collection, transmission and processing sub-system. No updating procedure can provide good forecasts if it is not possible to assess the value of the "true" error of the forecast, because the observed data are of an uncertainty that entirely overshadows the forecasting model's errors. No mathematical procedure, however sophisticated it may be and often is, can replace reliable data. It is however much easier to develop sophisticated computer programs than to manage a good data collection network in the field. This explains the proliferation of techniques and publications on hydrological forecasting, which are exclusively describing the mathematical operations and models, which can be easily included in the computer programs, but cannot produce reliable forecasts without improvement of the quality of the data.

4.3 EVALUATION OF HYDROLOGICAL FORECASTS

The evaluation or verification of hydrological forecasts is an activity which can be performed, unlike forecast updating, only after a rather long period of actual operations of a HFS by a well-established forecasting service. We will therefore limit this text to the general recommendations, based on WMO (1983).

A forecasting method can be held to be effective and its application justified only if the forecasting errors are substantially less than those which may result from a "naive" forecast. The mean value is perhaps the simplest such "naive" forecast although others are possible such as an assumption that the forecast variable will remain the same as observed at the initial moment of the forecast (termed inertial forecast), or a trend will continue into the immediate future.

The statistical evaluation of the effectiveness of any forecasting method should be based on a comparison with the performance of one of these "naive" forecasts. The coefficient of determination I_d measures the effectiveness of the forecast compared with the use of the mean.

$$I_d = 1 - \frac{s_e^2}{s_y^2}$$

where s_e is the standard error of the method, and s_y is the standard deviation of the forecast variable. A comparison with an inertial forecast is made by setting s_y to the standard error of inertial prediction.

A statistical parameter similar to the coefficient of determination can be used to evaluate the effectiveness of certain short-range forecasting methods.

$$I_\Delta = 1 - \frac{s_e^2}{s_\Delta^2}$$

where s_Δ is the mean square deviation of the forecast variable from its mean change during the forecast lead time. The value is given by the formula:

$$s_\Delta^2 = \sum_i^N (\Delta y_i - \overline{\Delta y})^2/N$$

Here $\Delta y = y_i - y_{i+1}$ is the change of forecast variable during the forecast lead time t (that is an error of of the inertial prediction) and $\overline{\Delta y}$ is the mean change.

The s_e,s_y and s_Δ values should be computed, taking into consideration their seasonal variations. The I_d and I_Δ coefficients are fully reliable statistical characteristics of both variables and errors in the case of a normal distribution.

When the distribution is non-normal, I_d and I_Δ can be used only after the similarity of the standardized error empirical distributions of both variables (the forecast and the reference method) is confirmed. The standardized error value expression is

$$z = (x-\bar{x})/s_x$$

where: z-is the standardized variable; x-is the random variable in question; $\bar{x}$-is the mean value of x; and s_x is the standard deviation of x. After standardization the mean value of a variable becomes zero and the standard deviation is unity. The similarity of the standardized error empirical distribution to the normal distribution should be checked using the Kolmogorov-Smirnov or Chi-square tests.

If a significant difference between both distributions is found, then the I_d and I_Δ coefficients provide only a rough estimation of the true forecast effectiveness. The I_d and I_Δ indices define the reliability of the forecast method in the following manner:

- if $I_d(I_\Delta) = 1$, the method is perfect and error-free
- if $I_d(I_\Delta) = 0$, the method does not provide any additional information over the "naive" forecast
- if $I_d(I_\Delta) < 0$, the method is generating greater errors than the "naive" forecast

Bearing in mind the probable errors in $I_d(I_\Delta)$, these indices must satisfy the following inequality for the forecast method to have any value.

$$\max(I_d, I_\Delta) \geq 1cr^{N}$$

where $1cr^{N}$ --critical value of the indices dependent on N the total number of uncorrelated forecast errors taken for the I_d or I_Δ computations. Critical values of 1cr are given below:

N	5	10	20	50	100
$1cr^{N}$	0,870	0,648	0,497	0,375	0,316

The criterion means that with a confidence of 95 percent the tested forecast method ensures at least a 10 percent reduction of the mean square error of the forecast with respect to the standard deviation of a forecasted variable.

A number of verfication criteria were evolved by WMO during the Intercomparison of Conceptual Models Used in Hydrological Forecasting Project carried out in 1972-1974. Various additional graphical and numerical verification criteria were used, including the introduction of a phasing coefficient and a coefficient of persistence to check for bias in the timing of the forecast hydrograph (see section 3.1).

A recently introduced concept has permitted the expression of the timeliness of forecasts and their accuracy as a single numerical score, expressed as units of time, and termed "mean forecast lead time" (MFLT). The score enables the effect of a particular combination of timeliness and accuracy on the recipient of the forecast to be examined. The concept of MFLT gives a measure of forecast effectiveness which is physically meaningful, and which is more closely related to economic benefit than are the verification criteria given earlier.

The calculation of errors of each forecast is essential for the accumulation of statistical data and for checking the stability of previous evaluations of the accuracy and effectiveness of a forecast method. This is particularly so when the method is of the type which might be developed from a relatively small amount of hydrological data.

Strictly speaking, statistical estimation of the success of an individual forecast is impossible, since the forecasting method itself only enables us to determine the confidence interval within which the appearance of the variable is expected with a given probability.

Hence, a forecast variable outside the confidence interval limits does not, as regards the theory of probability, mean that the forecast is unsatisfactory. The latter expression can be applied only to forecasts in which the value of the forecast variable is outside the limits of an interval having a confidence probability of almost 100 percent. This points to an error in the determination of the confidence interval itself.

The above evaluation and verification procedures are applicable in general to water level and flow forecasts in rivers and, in particular to flood forecasting. As indicated in section 1.4, there is a number of other hydrological forecasting activities, such as long-term water supply forecasting, specific low-flow forecasts, snowmelt forecasts, water quality forecasts, forecast of ice formation and break-up, lake level forecasts as well as flash-flood warnings. The reader is directed to the specialized national or international literature, in particular to Chapter 6 (Volume 2) of the WMO Guide to Hydrological Practices (WMO, 1983).

5. Benefit and cost analysis of hydrological forecasts

5.1 COST-BENEFIT ANALYSIS

In section 1.1 are enumerated reasons for establishing a HFS for different purposes, specified in section 1.4. Indeed the best reason for the establishment of such systems, particularly in developing countries, are examples of their benefits derived either in developed, industrialized countries or in neighbouring developing countries with similar climatic conditions. It should however be noted, that the benefits are not always the same and thus in a more careful analysis, during periods of national financial stringencies, the question will necessarily arise, if an HFS is an endeavour worth the investment. In the case of specific climatic conditions, such as those existing in regions subjected to tropical cyclones (typhoons, hurricanes), the justification of establishment of flood forecasting systems needs hardly a substantiation. Indeed the consideration of the following numbers will probably be sufficient.

It has been reckoned that, in the hundred years between 1870 and 1970, tropical cyclones, and the flooding which accompanies them, caused an average $1,500 million worth of damage and more than 5,000 deaths per year. Out of the world total the countries of Asia and the Far East suffered losses to the sum of $950 million a year, the countries of the Caribbean and the United States about $400 million, and the countries on the south-western shores of the Indian Ocean about $46 million per year. Again in Asia and the Far East alone, it is believed that river floods damaged or destroyed annually about 10 million acres of land and crops and affected the lives and well-being of over 17 million persons.

Yet in regions wher the climatic conditions are not as severe with respect to generation of floods as in the tropical cyclones affected areas, the decision-makers do often require a cost-benefit analysis, before the investment in a HFS is approved.

The determination of a realistic cost-benefit ratio for a particular application of hydrological forecasting is often a difficult and time-consuming task. Concepts of hydraulics, hydrology and economics have to be combined to develop comprehensive basin-analysis procedures. The essence of such procedures may be stated simply as:

(a) Evaluate the economic consequences of man's presence in the river basin without any forecasting service;

(b) Evaluate the same economic consequences of man's presence when a forecast service is provided;

(c) On the basis of steps (a) and (b), evaluate the benefits versus costs of a forecasting service, taking into account the reliability of the hydrological analysis.

Difficulties in evaluating intangible benefits and costs, especially those associated with the potential loss of life, continue to prevent their inclusion. However, in case studies undertaken to determine the cost-benefit ratio for flood forecasting, the results have shown an overwhelming margin of benefit over cost.

This general procedure is however different in the case of different kind of forecasts and different sectors of economy making use of them. The most frequent need for evaluation of HFS from a cost-benefit point occurs when the investment is to be made directly by the user, which can be a community, a power generating company, an agricultural complex, etc. and not necesssarily a government. This occurs predominantly in developed industrialized countries. It often consists in the evaluation of operations of a reservoir or a system of reservoirs.

In the case of hydroelectric stations with seasonal or carryover storage, flow forecasts will enable reservoir operating rules to be established and enable a suitable distribution of secondary energy generation to be scheduled during the drawdown period. This will maximize benefits by providing maximum kilowatt generation within the limitations of plan capacity. In terms of revenue it is advantageous to make predictions of generation and thus enable surplus energy to be sold at secondary rather than dump energy rates. However, if the reservoir operator bases his estimates on the lower confidence limit of predicted flows the use of forecast runoff in reservoir operation may be more restrictive than the use of rule curves based on historical records. For integrated hydro-thermal power systems with sufficient thermal reserve the question of confidence limits may be ignored. In many Canadian basins for example, more than 75% of the spring and summer flows result from snowmelt so that runoff volumes may be predicted with errors in the range of 20% - 30%.

It is of interest to note the conclusions from a study, "Benefits from Forecasting", by the Saskatchewan Water Resources Commission, which compared the value of hydroelectric generation at the Coteau Creek and Squaw Rapids plants of the Saskatchewan Power Commission with and without the benefit of long term (six months) river supply forecasts. The study showed an average annual forecasting benefit of $86,000 based on 38 years of records.

The ultimate improvement available from long term or seasonal forecasting is a function of the percentage of average annual runoff that is spilled from a reservoir or at downstream hydroelectric and irrigation schemes. Water wastage is also governed by the water usage capability of the project involved (i.e. peak turbine capacity). Because the net advantage available from long term flow forecasting is limited by the degree of regulation of a river, and by the size of installation, it is necessary to complete a detailed investigation of possible benefits for each scheme before additional hydrometeorological stations are installed for forecasting purposes. In the case of irrigation projects the long term or seasonal forecast may be used to govern the acreage planted each year which again provides a measurable economic value. Water supply projects for municipal or industrial use generally only consider firm or "guaranteed" flows, and hence, except in the case of marginal situations where demand exceeds the firm supply which may occur prior to a planned increase in water supply, long-term forecasting is only of limited value.

Long-term forecasting can be of significant value in the event of critical dry periods. The reservoir operator may curtail supply over a period of several months, providing a small current deficit so as to decrease the probability of a more severe water or energy shortage later in the drawdown-refill cycle.

The improvement of reservoir operations by long-term forecasting may also be of significant value in reducing concentration levels of pollutants. A restrictive or conservative reservoir operations mode designed to guarantee firm flows may result in water being stored unnecessarily which in turn may reduce dilutions below critical or acceptable levels.

Estimates of runoff volumes during normal or routine floods provide benefits in terms of increased water utilization. Reliable forecasts of flood volume enable the operator to use additional flow before the flood, and still ensure that the reservoir will rise to the level given by the operating rule curve. Sillage may be reduced or avoided. While the benefit in terms of increased water utilization may be evaluated directly, there is a secondary benefit as the operator may make scheduled releases prior to the flood peak and thus enable the routing effect of the reservoir to be fully utilized for flood mitigation. It should be remembered that with a gated spillway the peak flood releases may exceed those from the natural flood. Provision of an estimate of flood volume will encourage the operator to attain adequate drawdown prior to the onset of the flood.

Forecasts of unusual floods, or short-term (weekly) forecasts of river flows during flood periods may enable the movement of stock and equipment to higher ground to be completed or allow time for temperary protection to be raised which will reduced flood losses below historic levels.

5.2 COST-EFFICIENCY EVALUATION

While cost-benefit analysis may be an important source of information in the decision-making process in developed countries, this is not and should not be the case in developing countries, where the intangible benefits accruing from the prevention of loss of human life and human suffering are of paramount importance and cannot be quantified to obtain a meaningful cost-benefit ratio. For developing countries the concept of cost-effectiveness ratio, particularly for floods and droughts, should instead be the main criteria used for evaluating the design of the forecasting system.

By contrast to industrial countries developing countries usually have low population densities and low installed value of property in the flood plain and consequently structural flood defenses cannot be justified. Flood forecasting therefore provides the best way of reducing losses particularly those of human lives.

The characteristics outlined above give an indication of the potential benefits. In the case of developing countries except for flood forecasting no other means are available to protect life and property.

Whilst in absolute financial terms, the value of some of the property may seem low, in relative terms, expressed as the man-hours of work that have gone into obtaining it, its value is considerable. To replace it would require a considerable effort with consequent loss of other forms of productive output.

It is however perhaps in terms of river use that the maximum benefit comes. In much of the developing world rainfall is highly seasonal, a few months with rain and the rest of the year without. Where river flows are either perennial or with a flood peak out of phase with rains their agricultural potential is immense. A typical development would be a riparian rice scheme which is gravity fed. For this crop there is a critical period between sowing and replanting the seedlings; the period is around 26 days depending on the variety. Yields drop off at a rate of around 5% per day for early or late transplanting. The ability to give reasonable forecasts this far ahead enables seeding and transplanting schedules to be fixed to optimize the use of land as it becomes flooded.

It appears that while in absolute terms of cost-benefit ratio cannot justify the investment into forecasting systems in developing countries, in relative terms of pro-capita income per hour of work and prevention of suffering of the population, these systems are amply justified. What is however more difficult to justify, is the unnecessary luxury with which some consulting firms are designing HFS in developing countries, often not as projects of direct technical assistance but as regular investments paid by soft-loans offered by the governments of industrial countries. Indeed, as was demonstrated in several parts of this text, relatively simple data collection systems and models provide forecasts of an accuracy, which is sufficient if compared with the level of accuracy of measurment of historical and real-time data. Thus proposals of more sophisticated systems, particularly with very complex equipment and forecasting procedures requiring sometimes an order of magnitude of higher investment in the hardware and software are being offered. Instead of a simple system costing about US$200,000, a system costing around US$2 million was recently proposed to a relatively very poor country of Asia. A radar, costing by itself about US$1 million is frequently proposed as the only panacea for flood control. Yet with the low qualification and scarcity of trained personnel, there is no chance to increase the efficiency of such sophisticated systems as compared with a relatively simple and low-costing ones. Quite to the contrary, since the sophisticated hardware and software is much more difficult to maintain and adapt to new data and needs, the developing country is sometimes left with a non-operating sophisticated system. The higher cost in the end not only does not produce a higher efficiency of the system, but may cause it to become zero. A repayment of the loan becomes a burden to the developing country without any benefit. In such cases the advice of objective and qualified specialists, without any vested interest in the cost of the proposed HFS should be sought.

6. Examples of design of HFS in WMO-assisted projects

6.1 NETWORK DESIGN

Guinea-Mali, Niger Basin

This system was designed in 1969 and commissioned in 1974. The rivers concerned were the Niger itself from its source to its confluence with the River Bani in Mali, and its main tributaries, the Milo and Tinkisso in Guinea and the Sankarani and Bani in Mali. A total of 77 hydrometric and rainfall stations were taken over by the project and available for the calibration of the forecasting models. At the beginning of the project, 12 stations were equipped with radio-transmitters and could be used for real-time data transmission, and thus for operational forecasts. The size of the basin and its physiography, consisting of the high mountains of Fouta Djallon in the upper reaches of the rivers and of flat valleys entering the arid plain of the middle reach after the confluence of the Niger with the Bani (the inner delta of the Niger), required more detailed real-time information on the flows and rainfall. For this reason 15 additional radio stations were established, each to transmit both rainfall and discharge data daily during the rainy and flood season, including the period of flood recession. The total of 27 radio stations were installed so that each larger tributary could be monitored by a discharge station, either before or after its confluence with a river of second or first order (Niger).

After the commissioning of the sytem and gradual decay of the radio transmitting facilities, it became obvious that the forecasts did not lose much of their accuracy with the decrease in available real-time data. An analysis of the system indicated that the original network design over-estimated the need for real-time reporting stations. The old system has now been superseded by a new HFS which, at present, is intended to provide flood-routing forecasts only. The number of real-time reporting stations has been established at 95 for the entire Niger River and its main tributaries, in particular the Benoué, from its source in Guinea to its estuary in Nigeria. Due to funding constraints, only 65 of these stations are being supplied with real-time radio transmitting, facilities. Figures 19 and 20 represent the networks of the former and the present forecasting systems.

Pampanga River (Philippines)

This HFS was established on the recommendation of WMO (made in 1971) with the bilateral assistance of Japan (see Annex 3). The 8550 km^2 basin is very mountainous. A total of 11 stations reporting in real-time were selected, principally on the basis of their suitability for VHF radio transmission after field propagation tests. Three stations report only rainfall, eight stations report

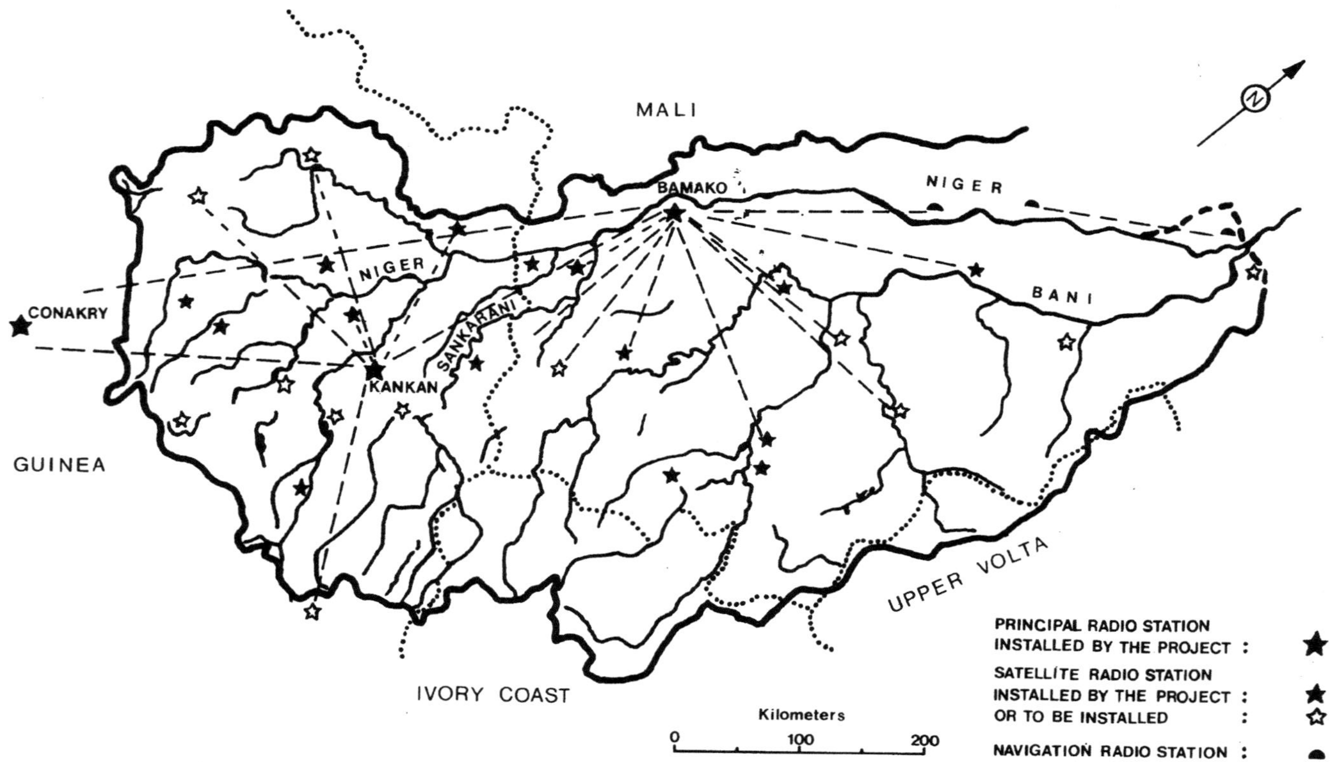

Figure 19. Flood warning system on the River Niger (Guinee, Mali)

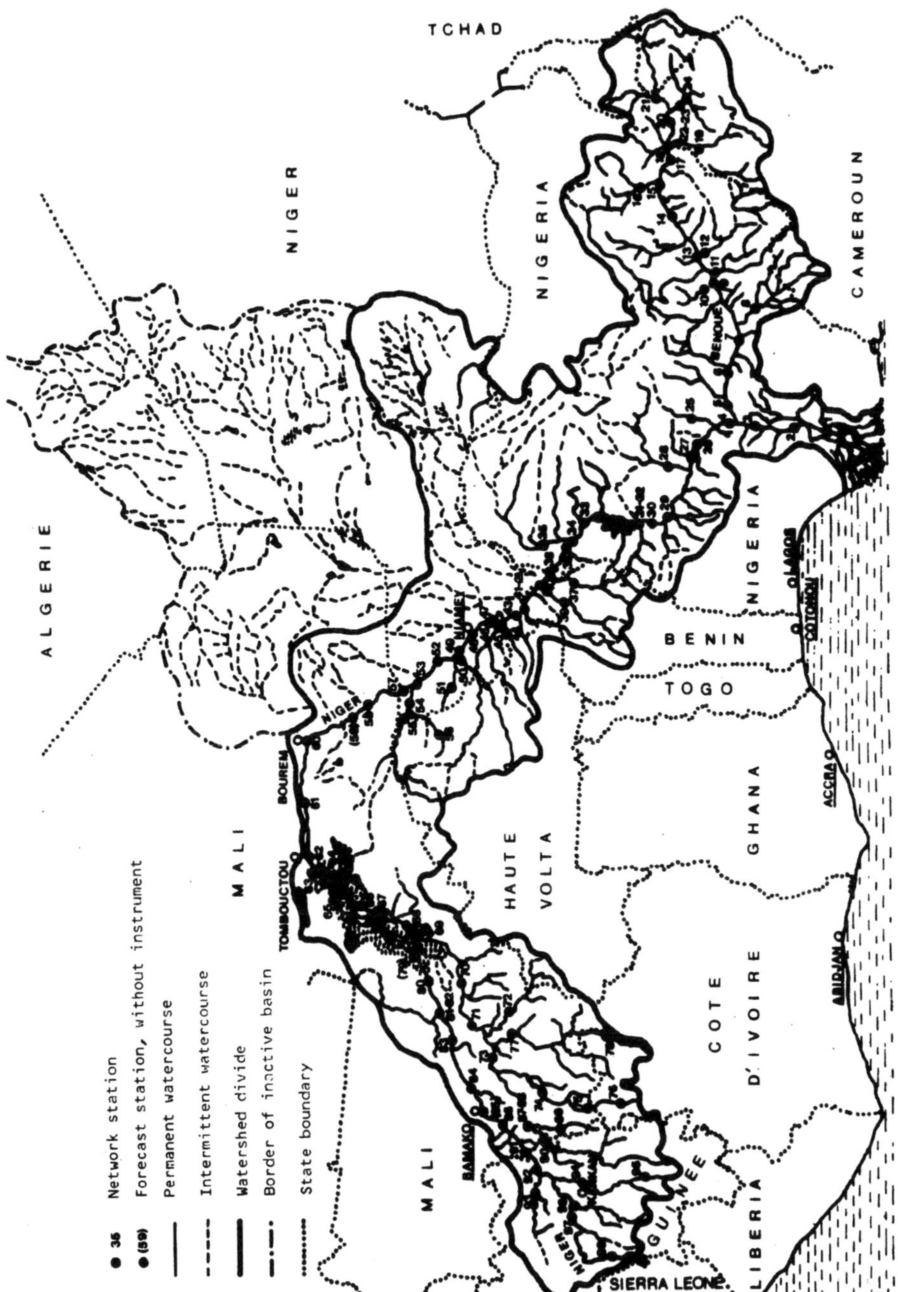

Figure 20. Network of stations in the forecasting system for the whole River Niger.

both rainfall and streamflow. The system functions almost exclusively during the typhoon season. It was commissioned in 1975.

Indus River (Pakistan)

The system is described in Annex 4. The network was designed in 1977. The river is formed by five main tributaries in the Punjab: the Indus, the Jhelum, the Chenab, the Ravi and the Sutlej. All the upstream reaches of these rivers are in the Himalayas, where only very few possibilities for observation exist, either because of difficult access or due to location in territories which are disputed by two countries. Once in the Punjab plains, the river flows for over 1000 km over relatively flat terrain, where routing procedures require predominantly discharge stations. To overcome the inaccessibility of the upper reaches with respect to rainfall data acquisition, a quantitative rainfall measurement radar was installed in the foothills of the Himalayas. The network's design was based on the existing meteorological network serving irrigation planning and air transport, and the hydrometric network serving both irrigation and hydropower production. The so-called "rim-stations" situated at the foot of the mountains, where the rivers enter the large plains, had relatively long-term observations and were used as basic stations for calibrating the forecasting procedures. In designing the network of streamflow stations, it was ascertained that the major rivers were well gauged but there was insufficient gauging of the smaller tributaries. A total of 24 additional rainfall stations was proposed of which 19 were near rivers on which streamgauging stations were also proposed. The real-time reporting network was finalized with the following configuration: 36 VHF radio-telemetered stations with rainfall tipping-bucket gauges, of which 31 were connected to water-level recorders at streamgauging stations; five HF (voice SSB-radio) stations with both rainfall and water-level recorders; and 11 additional stations reporting in real time via radio-communication systems of other governmental authorities (the police in particular). This configuration still leaves much to be desired with respect to snow-depth reporting and rainfall and discharge measurement in the upper reaches in the mountains. The shortcomings of the network were reflected in the accuracy of the forecasts, although the models were adapted to the lack of information from some parts of the basin by using extrapolation techniques. It should be noted that the real-time reporting network requires 13 repeaters and serves five data-receiving centres, of which only two make system-wide forecasts.

The Amazon River in Brazil

When WMO, together with the UNDP Office of Project Execution, was entrusted with the execution of the project on improvement of the knowledge of the hydrology and climatology in the basin of the River Amazon in Brazil (the river basin covers territory belonging to seven countries), the establishment of a HFS in certain parts of the basin, although a component of the project, was only a marginal objective. More emphasis was given to the planification and installation of a system of observations (networks) and related processing of data to assess the potential of the basin for its economic development. Thus, in parts of the basin, network design was a more important objective than was forecasting of flows. The HFS installed are described in Annex 5. The number of streamgauging stations in a basin of some $3.5.10^6$ km^2 was 169 in the Amazon basin proper (operated by five different agencies).

There were an additional 91 stations in the Tocantins-Araguaia basin, in which a hydropower development required improvement of forecasts. Some of these stations measured levels only. In 1979, for all practical purposes, streamflow data were available from 90 stations in the Amazon basin proper and from about 65 in the Tocantins-Araguaia basin. The corresponding density was thus about 40,000 km^2/station for the Amazon, 5,000 km^2/station for the Tocantins. There were 341 meteorological stations in the basin, of which 234 measured only rainfall. Ninety-nine stations were equipped with a recorder. The distribution of the stations was very unequal, in particular over the western and northern parts of the basin (except for the Upper Rio Branco) where the mountainous regions were practically without stations.

The cost of operating stations under conditions of difficult access is exorbitant. In 1976, the average annual cost of operating a streamgauging station was US$25-50,000 per station. In the Tocantins basin the cost was lower, but was still about US$10,000 per year/station.

A complex exercise of network design-cum-data processing, based on the so-called "square-grid technique" was undertaken, but the results have yet to be evaluated. As a target resulting from the delineation of seven homogeneous regions, the number of streamgauging stations should be increased to 400-500, and that of meteorological (precipitation) stations to about 1400. This network is to serve to compile an inventory of water resources. It is obvious that calibration of models for forecasting purposes will have to be based either on already existing stations or on the newly established ones, after they have been in operation for a few years.

6.2 DATA-COLLECTION SUB-SYSTEMS (SENSORS AND TRANSMISSION)

Niger River basin

The present HFS in the Niger River Basin is presented in Annex 6. It is however considered useful to describe some stages of its design, in which the author participated most actively.

In systems where telemetering is not automatic and thus an on-line sensor and transmitter are not necessary - most frequently the case with observers who telephone, telex or radio their observations to the forecasting centre, the sensor is almost entirely independent of a telecommunication link. In this case the choice of the sensor depends only on standardization of equipment and economic constraints. This was the case with the original HFS on the upper Niger River in Guinea and Mali(see para 6.1.1). The raingauges used were of several types (mostly tipping-bucket, French-made "Précis"). The stage recorders were either of the float type (OH, SIAPE and Neyrpic - later Neyrtec) or of the bubble-gauge type (Neyrpic, Seba). In the present project, covering the entire Niger Basin, where the transmission of the data from 65 (later from 95) stations is to be provided by the TIROS-N polar-orbiting satellite through the ARGOS system, the choice of sensors was carefully standardized in order to avoid the problems which occurred in the design of the HFS on the River Indus in Pakistan, as reported below. Indeed, the coding data for onward transmission by DCPs (Data Collection Platforms) to the satellite, for retransmission to the individual national and the international forecasting centres, requires a sensor-DCP interface. The selection was made on the basis of both technical and economic parameters, with the data-transmission sub-system selection preceding that of the sensor selection. A specification of requirements was prepared (see Annex 6) and seven companies participated in the bidding.

In five of the bids the Argos system was proposed. This is an operational system using two of the polar-orbiting TIROS-N satellites. It has been working since 1978 and is guaranteed until the end of the current decade. A failing satellite can be immediately replaced. Even with one satellite in operation it can fulfill the specification as to the frequency of data transmission. At present no fees have to be paid when the data user has a receiving station and thus does not require the services of the data centre in Toulouse. The Argos system is suitable for a data-transmission system for the River Niger Basin.

In three of the bids, the Meteosat (European Geostationary Satellite) data-collection system was proposed. The principle of the Meteosat system is suitable for a hydrological forecasting system as the stations can transmit data at regular intervals, when a particular parameter has been exceeded or upon a signal from the satellite. Meteosat I data collection has been working since 1977. However, the Meteosat system was not regarded as an operational system on account of the malfunctioning of the second satellite. Data collection has not been achieved with this satellite and a decision on the operational phase had not been taken at the time. Due to these uncertainties and to the fact that the equiment of the Argos system is less costly than that of Meteosat, the Argos system was considered to be the most suitable of the two satellite-based systems. About one year after the decision on the selection of the system, an intergovernmental agreement was reached to make Meteosat operational.

As to other types of transmission systems, one company offered meteorburst communication and another classical UHF/VHF radio (ground-based) transmission. These systems have the advantage of independent ownership and do not rely on any satellite.

Proposed transmission systems

COMPANY	ARGOS	METEOSAT (Geostationary)	METEOBURST	UHF/VHF (ground-based)
A (Europe)	X	X		
B (Europe)	X	X		
C (USA)			X	
D (Japan)	X			
E (Europe)	X			
F (Europe)		X		X 2 alt.
G (USA)	X			

The system, as it functions today is described in Annex 6.

Indus River (Pakistan)

The sensors for this project were ordered ahead of time in order to allow a scheduled installation of the network, in particular of the water-level recorders. However, when the data-transmission system was being selected, it appeared that the sensors which had already been delivered could not be interfaced with DCPs without additional adaptations, which were more expensive than the purchase of new sensors, adapted and provided by the manufacturer with the DCPs. Fortunately, the hydrological services of Pakistan needed sensors of the type originally purchased for other networks, not used in the HFS, and thus no loss resulted from this situation.

The selection of the transmission system was limited to systems via polar-orbiting satellite, land-based HF, VHF or UHF systems, and the Meteorburst system. There was then and still is no geostationary satellite in orbit above this region available for data transmission. In view of very closed mountain valleys, the polar-orbiting satellite sytem could present problems. A similar bidding process as for the Niger River basin system was undertaken (in fact, the design of the Indus River system preceded that of the River Niger). The Government of Pakistan decided in favour of a land-based VHF system supplied by a company from the USA. The communications are by VHF radio within the frequency range 148-171 MHz. It has the following four elements:

(a) Sensor and data transmitter
(b) Repeaters
(c) Central receiver
(d) Data display terminal

Basic principle

When a change occurs in the river level or rain falls at the station, this new value is stored within the data transmitter. This value, along with the station identification number, is then transmitted to the central station. VHF radio waves travel in a straight line rather than following the Earth's curvature. Thus repeaters are required. Because of the long distances involved in this system, most of the weather data have to pass through one or more repeaters before reaching the Flood Forecasting Centre in Lahore, where the data are received by a 60 m high antenna. The data display terminal takes the data from the receiver(s), updates the files for the identified station and displays the previous and the updated data on the printer and map display. Simultaneously, the data are made available to the NOVA computer for processing and forecasting.

Sensors and data transmitter

Two types of sensors are installed: tipping-bucket raingauge and float-type river-level gauge (other sensors can be added). The raingauge has a precision of one millimeter. The water-level gauge was originally designed for a precision of one centimeter but, due to uncontrollable oscillations of the sensor and resulting battery drain from unduly frequent transmissions, this precision was extended to 3 cm.

The data transmitter is completely enclosed. Each data sensor has its own identification number. When a sensor detects a change in river or rain level, the corresponding accumulator circuit in the logic board is incremented or decremented. The data report, consisting of the new value in the accumulator and the identification number, is immediately transmitted to the receiving station in Lahore and to the receiving stations at Mangla and Tarbela, if appropriate. In addition, a timer circuit causes the current value in the first accumulator of each DCP to be sent twice a day, as a verification that the station is in operating order. A crystal-controlled clock is used for driving all timing functions, including the baud rate clock. The battery needs to be recharged only once a year.

The VHF system consists of two paths (see Annex 4): one for the upstream part, the Ravi and Sutlej Rivers, and the other for the Jhelum, Chenab and Indus Rivers. The first path consists of eight stations reporting to Lahore through repeaters at Chunian, Nandipur and Pasrur, and two stations reporting directly to Lahore. In addition, QPM radar data are transmitted from Sialkot via the repeater at Nandipur. The second path consists of 26 stations reporting to Lahore through two repeater chains which join together at Chakwal before coming through Nandipur. One chain transfers data from the southern sector around Multan to Lahore. The other chain transfers data about hydromet conditions affecting Tarbela and Mangla reservoirs to Lahore, as well as to each of these dam sites. (Appropriate Mangla/Tarbela data are concurrently received on site to assist in reservoir operations for reducing potential of downstream flooding through synchronization of the peaks.)

Repeaters automatically receive data on one VHF frequency from either a DCP or another repeater and retransmit the same on a different VHF frequency to another repeater or to a cental station. A repeater consists of a receiving radio tuned to one specified frequency, a transmitting radio tuned to another specified frequency, and the repeater control board. Repeaters are always waiting for an incoming signal. When a signal of the proper frequency and format is received, the control board applies power to the transmitter, turning it on. Power for repeaters is supplied from batteries to ensure their continued operation even during power shortages. The source can be either: (i) two 12-volt batteries supplied by a low-flow trickle charger connected to an AC source or, (ii) two 12-volt batteries supplied by a solar-panel charger. The first method is utilized where AC power is readily available. The second is utilized at stations such as those on hill tops where no electric power is available.

The central station data receiver receives the data transmitted from DCPs or repeaters and converts it into digital form for direct input to the Nova computer or data display terminal. The receiver consists of a radio receiver and a signal conditioner. The radio receiver is identical to the one in the repeaters. VHF waves are converted into two audio tones which enter the signal conditioner. The two audio tones are demodulated into the digital data they represent. A check is maintained on the receiver to ensure that the signal is of proper format and frequency. The digital data is available at two output ports: one for Nova computer connection and the other for connections with the data display terminal.

The data display terminal was a microprocessor-based system used for receiving, processing, storing and displaying real-time hydromet data, since it was replaced by IBM microcomputers, which perform the functions described in part 2.3 in connexion with the real-time data base management system "Alert".

The system was installed under the supervision of the manufacturer, by local teams, and most of the infrastructure (standardized DCP antenna towers, etc.) was produced in Pakistan. The transmission system is maintained by one of the agencies involved in the system (WAPDA). A new project has been started recently to improve this system.

The radar is a supplement to the raingauge network and its value depends on its operational readiness. Due to its remote location (100 km from the forecasting centre) and sophisticated maintenance needs, it is unable to perform with necessary reliability and accuracy, despite its design characteristics. Tt is an IC-band (3.36 cm), narrow-beam width (0.9°) giving the best possible permanent echo discrimination), computer-controlled, intalled at Sialkot. The radar has been programmed to measure average rainfall in 17 sub-catchments/areas. These areas have been specifically designed for input to rainfall-runoff models.

The data of the 17 sub-catchments for flood-forecasting purposes is passed on to the Flood Forecasting and Warning Centre in Lahore for feeding into the Lahore computer (NOVA 3/12, 96 K) as input to the rainfall-runoff model.

Due to many uncertainties surrounding the use of radar for precipitation measurement, it is unlikely that an average accuracy exceeding a factor of 2 can be achieved from these radar measurements.

Amazon basin (Brazil)

The network-design exercise in this basin indicated that a real-time data transmission based on satellites or meteorburst is the only solution for the acquisition, at an acceptable cost, of the hydrological and climatological information, not only for forecasting purposes but in general. After evaluation of all possibilities, a pilot system based on the geostationary GOES satellite of the USA was selected and commissioned in 1983, covering the River Tocantins basin (75,000 km^2), as part of a FORTH system used for several purposes, but mainly for hydropower production (see Annex 5). The system uses ten DCPs with three to four sensors and one battery-check channel, for data on precipitation, water level, air temperature and air humidity. The data are relayed by the satellite to a central receiving station at Sao Jose dos Campos (near Sao Paulo), where the national space investigation research institute (INPE) is located. From there the processed data are relayed by telex to Brasilia and Belem. A computer-to-computer link assists in producing forecasts at the SUDAM office in Belem. The receiving station at Sao Jose dos Campos comprises: one satellite-synchronized clock, two pre-amplifiers, two downconverters, two receivers, two demodulators, two PDP 11/03 microprocessors, and one Data General Nova microcomputer (128 kb and magnetic tape drive) for "after reception" data processing and its distribution to users.

The installation of the system provided experience of use in its further expansion. It is most useful to have a mechanical and electrical workshop at hand. The most important factor during installation of a DCP is availability of a communication link with the ground receiving station or an alternative (portable) link with the satellite, to permit the checking of the correct functioning of the installed DCP. For the USA GOES, such a link is provided by a portable unit connecting with the GOES "Data Acquisition and Monitoring Sub-system" (DAMS), providing messages on the quality of the transmission of the DCP being installed ("DAMS health messages"). This instrument provides the installation crew with confidence in the performance of the installed station and allows for adjustments to be made before the crew leave the site.

The DCPs collect data from the sensors every half hour and transmit to the satellite every three hours - at one minute intervals between each DCP. The message contains all data collected during the past six hours (thus also data already transmitted three hours earlier) to permit a back-up, should one transmission be lost. The data are received by the ground station and stored in the microcomputer. Through an interface with a telex, the users can call up at any time to receive the data on their own telex machines or through a computer-to-computer link. The system's DCPs are maintained, for the time being, on a bi-monthly schedule, but the operational maintenance is foreseen as a six-monthly routine. Emergency visits to two DCPs have been necessary so far. The present state of the system is considered satisfactory. For the expansion of the system and for its full use, close co-operation with the US operator of the satellite (NESDIS - National Environmental Satellite Data and Information Services) appears a most important prerequisite. Similar equipment is contemplated for installation in a system in Panama.

Ganges River (Bangladesh)

The HFS proposed and implemented over the past three years uses conventional sensors and voice HF radio transceivers (single-side band). Twenty-four sets are in operation and a further 30 are being installed. In this case no particular interface with the sensors is necessary. The central station is another transceiver of the SSB type. Reliability considerations and mainly the possibility of human error on both sides of the telecommunication channel have led the Government to consider a VHF land-based telemetry system. A pilot project, consisting of four water-level recorders and two raingauges with interrogated transceivers, supplied through a bilateral assistance donor, is being intalled for testing purposes.

Yangtze River basin (China)

A simple system for data collection, as used in Pakistan on the Indus River, but for very small catchments, was used in two pilot projects using HFS in the People's Republic of China, on the Yellow and Yangtze River basins. The Yangtze River basin system, in the basin of the Pu Che tributary, is described in Annex 7.

6.3 DATA-PROCESSING AND MODELLING SUB-SYSTEMS

Indus River (Pakistan)

During the design of the FORTH system in this river basin three levels of complexity (A, B, C) of forecasting procedures (models) were matched to three computing facilities (computers) available or purchased for the project. One of the government agencies responsible for the project (WAPDA) was able to provide time on a larger main-frame computer (IBM 370). A minicomputer, identical to that used with the rainfall-measuring radar NOVA 3/12, and a desktop programmable calculator HP9825A, were purchased with the assistance of the WMO/UNDP project. Two types of models were developed by sub-contractors: for the basins above the rim-stations (at the foot of the mountains), two rainfall/runoff models, the CLS and the API Continuous (see para. 2.5.3); for the routing of the flow from below the rim stations to the estuary below Karachi, a hydrologic routing model was calibrated. This is a non-linear cascade of reservoirs. A hydraulic routing model based on the full differential equation of unsteady flow was calibrated in level A for those reaches of the Rivers Ravi and Chenab which had sufficient topographical data on the channels and some roughness coefficient measurements. Both in development (calibration) mode and operational (forecasting) mode, the level B models (CLS, APIC and the non-linear cascade routing) were installed on the NOVA 3/12 computer. Level C, designed for back-up of these same models in case of a power cut or computer breakdown, was run only in operational mode on the HP 825A desktop calculator, using rechargable batteries. The reliability of level B was enhanced by the fact that the NOVA 3/12 computer in the Flood Forecasting Centre in Lahore and that in the radar station in Sialkot were identical and their spare parts were interchangable. Unfortunately, however, this type of computer, supplied as a package by the radar company, proved to be subject to frequent breakdowns. Nevertheless, since the commissioning of the system in 1980 up to 1983, with yearly maintenance, the models on levels B and C have provided forecasts with an accuracy as expected. The main problem in ascertaining the accuracy of the forecast flows appeared to be paradoxically the "true" values of the observed flows which, due to difficulties experienced in streamgauging, were subject to errors larger than those probable in the computation of the forecasts. Once again the observational data appeared as the alpha and omega for the success of a HFS.

Amazon basin (Brazil)

Most of the reaches on the Rivers Tocantins and Amazon were not forecast before the start of the WMO/UNDP project. Only some reaches of the River Tocantins, above the Tucurui dam, were modelled by the routing component of the SSARR (US Corps of Engineers) model with relatively unsatisfactory results. As a first step to improve forecasting on the reaches of the rivers in the Amazon basin of some economic importance, multiple correlation routing models (see para.2.5), using difference in stages rather than stages themselves, were developed by the WMO experts and consultants using available time on the SUDAM computers (Burroughs 2700 and a minicomputer Cobra 400). An HP 9845B and an HP 9831A desktop computers were purchased to improve the available computing capacity and provide insurance against breakdowns of the larger computers. The correlation models were programmed on the HP 9845B but it soon became clear that if any significant improvements were to be achieved, better computing facilities and better models would have to be provided. A bidding exercise provided for building both rainfall/runoff models for the middle reaches and routing models for the lower reaches by a sub-contractor from Argentina (see Figures 21 and 22).

The rainfall/runoff model (PLUMUS) is of the moisture-accounting type (see A.2.d in para. 2.5) and was calibrated for the middle reaches of the Rivers Madeira, Purus, Solimoes, Negro and Branco and a few "smaller" rivers flowing from intermediate catchments. The discharges of these rainfall/runoff models are inputs to hydrologic routing models (see A.1.b in para. 2.5) for the above rivers up to the profile of the main Amazon at Manaus, from where a hydraulic routing model (see A.1.c in para. 2.5) is used to route flows up to the gauging station Santarem below Obidos, the end of the back-water caused by the tidal effect (about 800 km from the estuary near Belem). Due to lack of reliable measurements of the back-water effect, the modelling of the reach Santarem-Belem is difficult if not impossible at this time. Consideration has been given to improving the computing facilities by interconnexion of the Belem SUDAM computers (Burroughs, Cobra, HP 9845B) with large mainframes in Brasilia via dedicated microwave telephone line but, due to many technical and institutional difficulties, the solution at present is focused on the purchase of a new computer in Belem.

Niger River

The first HFS on the Upper Niger used rainfall/runoff models of the multiple correlation type for the upper reaches (see A.2.a in para. 2.5) and hydrologic routing models of the linear cascade type (Kalinin-Milyukov)(see A.1.b in para.2.5). The models used a small desktop programmable Olivetti calculator. The new HFS being built for the entire river is to use two PDP 11/60 computers already available in the computing centre of another WMO/UNDP project (AGRHYMET) in Niamey (128 Kb, 2 dual flopy disks each of 1Mb, two

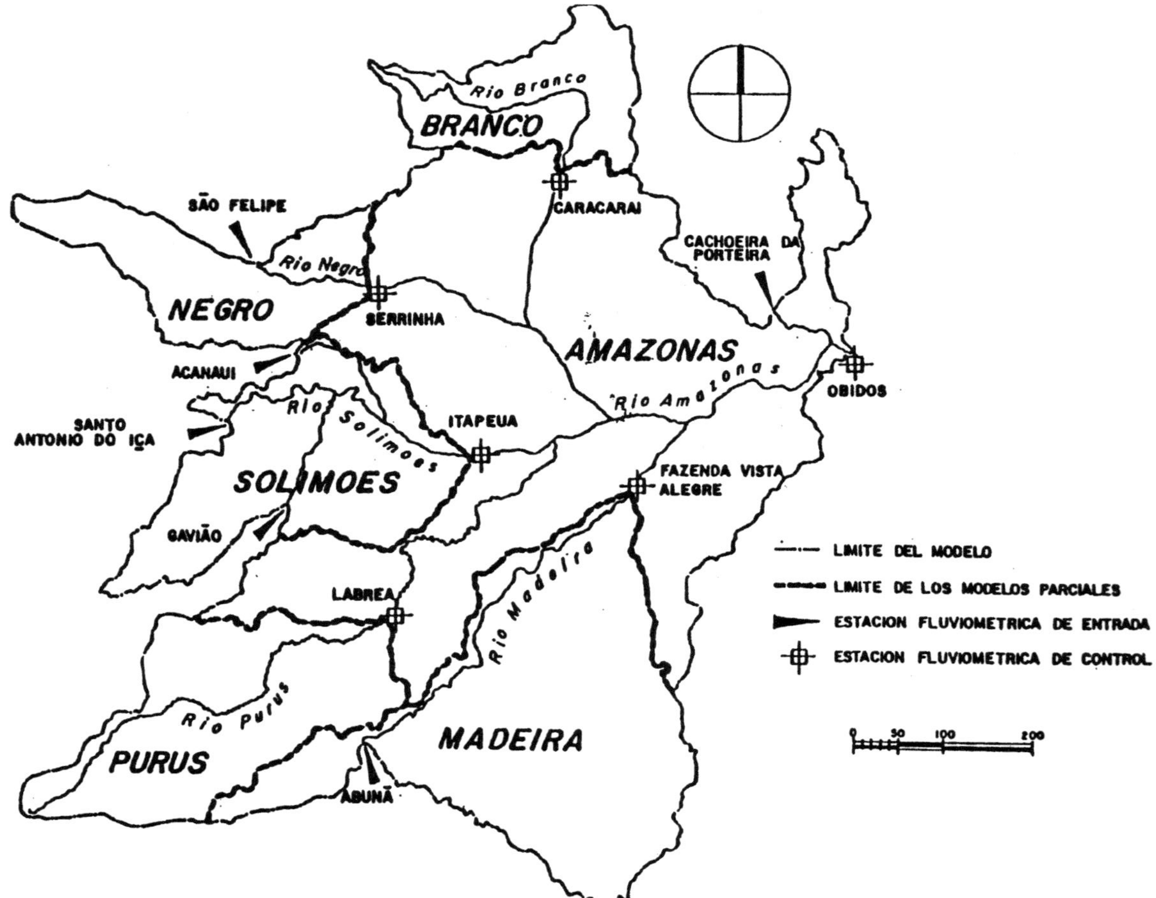

Figure 21. Hydrological (forecasting) models on the Lower Amazon River in Brazil.

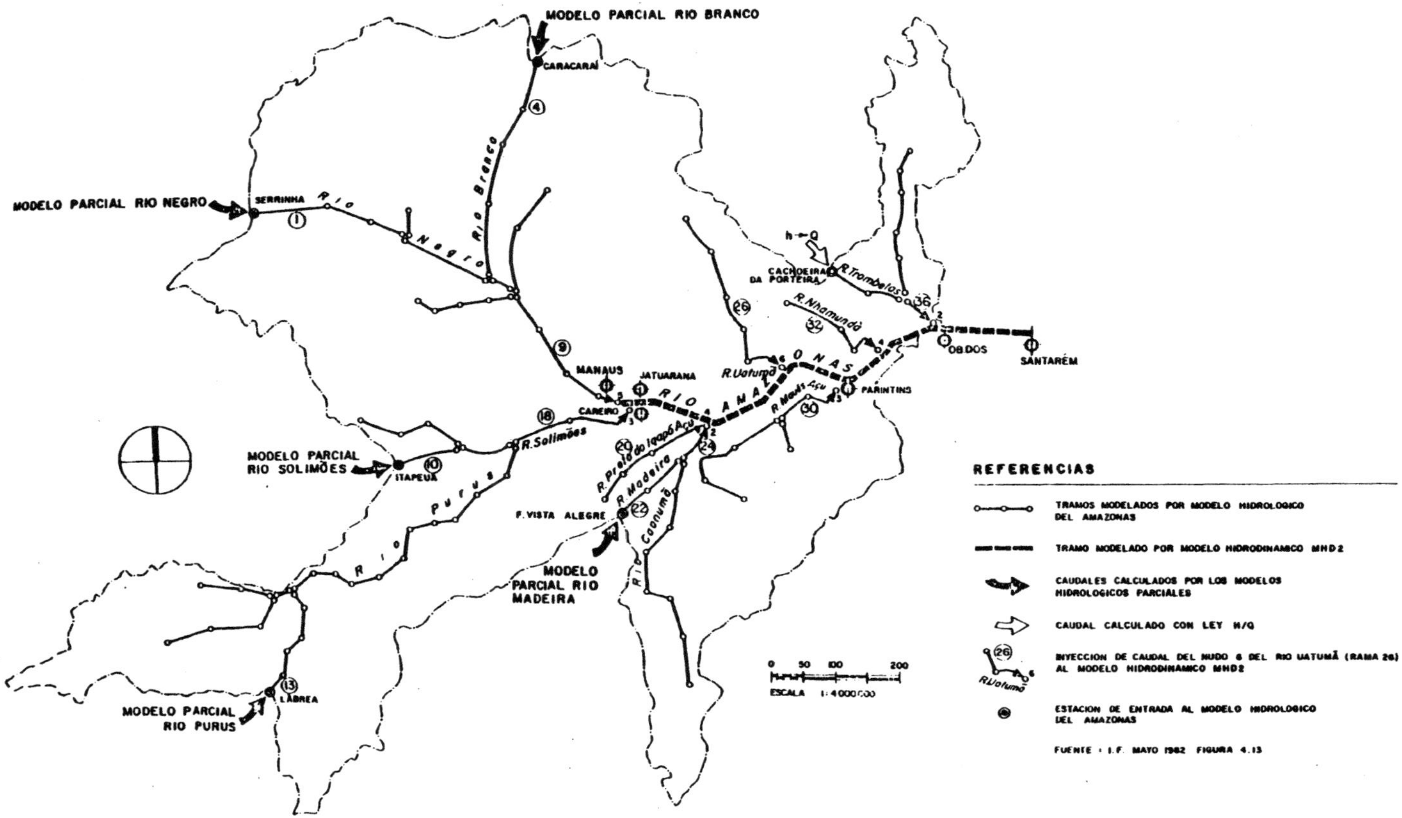

Figure 22. Hydrological and hydrodynamic forecasting models on the Amazon River in Brazil.

magnetic tape drives, 2 disk drives of 176 Mb each, removable, and one printer/plotter).

Citarum River (Indonesia)

The HFS installed on this Indonesian river was conceived as a pilot system to assist the Institute of Hydraulic Engineering in Bandung in achieving self-sufficiency in all aspects of hydrological forecasting. The system also had practical implications as it was to provide forecasts of inflow to the Jatiluhur reservoir, on which a hydropower plant is installed. The forecasts of flow below the reservoir were mainly for flood-warning purposes. The basin above the reservoir is 4,600 km^2. The data-collection sub-system consists of telephones, VHF and HF SSB radio voice transceivers, and a total of about 40 stations report precipitation and discharges. The mountainous basin presents large differences in rainfall depth. The basic model used in the FORTH system was SSARR (US Corps of Engineers) in its reduced form (COSSARR), as adapted by the Mekong River Basin Secretariat. The computer used was the CII Honeywell Bull Mini-6, already available at the Institute. During the WMO-assisted project, two other simpler models (linear systems approach) were calibrated in a comparison with the COSSARR. It was concluded that, under conditions lacking in data accuracy, the simpler models provided forecasts of similar accuracy to those made with the computationally and computer-time-exigent SSARR model. The conclusion of a report prepared by an independent consultant indicated that, while the COSSARR model has performed adequately during trial runs, a similar forecasting accuracy and hence a similar level of benefit can be obtained from a simpler model costing much less to develop, calibrate and run operationally.

6.4 FORECAST UPDATING AND DISSEMINATION

As indicated in para. 6.3, the biggest obstacle in using real-time streamgauging data for updating forecasts in most systems in developing countries assisted by WMO projects is the low accuracy of the streamgauging operations during floods. In the majority of cases, no parameters have been updated, but only simple adjustments made to the hydrographs and/or time-base. On the other hand, WMO has included plans for forecast dissemination and disaster prevention in all its assistance projects for the establishment of HFS. It should, however, be stressed that, within the family of international organizations, WMO is not the principal and certainly not the only one to assist developing countries in their plans for dissemination of warnings and disaster mitigation operations. In particular the United Nations Disaster Relief Office and the League of Red Cross Societies are the main organizations with which WMO co-operates in assisting developing countries in adding one of the most important sub-systems in the HFS: forecast dissemination and disaster mitigation. As an example of the procedures used in forecast dissemination and disaster mitigation in the case of floods, Annex 8 contains the Jhelum District flood order, prepared within the framework of WMO/UNDP assistance to the Government of Pakistan in establishing a HFS in the Indus River basin.

References

Attmanspacher, W. and Schultz, G.A. (1980): Möglichkeiten un potenzieller Nutzen eines Bundesdeutschen Niederschlagsradarverbund System, Wasserwitschaft, RFD.

Bissel, V.C., Pasteris, P.A. and Bennet, D.G. (1982): Standard Hydrologic Exchange Format (SHEF) for automated data transmission. Suabmitted to ASCE Journal of Water Resources Planning and Management Division.

Bobinski, Eryk (1979): Technical report on On-line (real-time) hydrological forecasting systems. A report to WMO Commission for Hydrology, WMO, Geneva.

Box, G.E.P. and Jenkins, G.M. (1970): Time series analysis: Forecasting and control. Holden and Day, San Francisco, Calif., USA, 553 P.

Burnash, R.J.C. (1982): Criteria for the selection of a computer for FORTH systems. Personal communication.

Burnash, R.J.C., Ferral, R.L. and McGuire, R.A. (1973): A general streamflow simulation system - conceptual modelling for digital computers. Report by the Joint Federal State River Forecast Center, Sacramento.

Burnash, R.J.C., Ferral, R.L. and Strem, E.T. (1981): Area evaluation of precipitation, a technique for maximizing operational information from a limited raingauge network. Report submitted to the annual meeting of the AMS.

Burnash, R.J.C. and Strem, E.T. (1983): Sensitivity of forecast made by the Sacramento model precipitation data in the River Eel basin. Personal communication.

Clift, G.A. (1985): "Use of radar in meteorology", Technical Note No. 181, Publication No. 625, WMO, Geneva, Switzerland

Dooge, J.C.I. (1973): Linear theory of hydrologic systems. Tech. Bull. No. 1468, US Dept. of Agriculture, ARS, Washington, D.C.

Gabris, P. (1985): "Problems of identification of hydrologic systems (with particular reference to deterministic forecasting models) (in Slovak), PhD thesis, Institute of Hydraulics and Hydrology, Slovak Academy of Sciences, Bratislava, Czechoslovakia.

Greene, D.R., Hudlow, M.D., Farnsworth, R.K. (1979): A multiple sensor rainfall analysis system, Third Conference on Hydrometeorology, AMS, Boston.

Guidelines for Computerized Data Processing in Operational Hydrology and Land and Water Management (1985): Joint FAO/WMO Publication, WMO-No. 634, Geneva.

Herschy, R.W. (1978): Accuracy, Ch. 10 in Hydrometry. John Wiley & Sons, Chichester, UK.

Hudlow, M.D., Farnsworth, R.K., Ahmert, P.R. (1984): NEXRAD Technical Requirements for Precipitation Estimation and Accompanying Economic Benefits; HYDRO Technical Note 4, Office of Hydrology, NWS-NOAA, Washington, USA.

Ishizaki, K. (1979): Areal assessment of precipitation. A report to the WMO Commission for Hydrology, WMO, Geneva.

Kagan, R.L. (1972): Planning the spatial distribution of hydro-meteorological stations to meet an error criterion. In: Casebook on hydrological network design practice, WMO Publ. No. 324, WMO, Geneva.

Keith, Beven (1985): Distributed Models; in "Hydrological Forecasting", edited by M.G. Anderson and T.P. Burt, John Wiley, Chichester, UK.

Klemes, V. (1982): The desirable degree of rigor in testing rainfall-runoff models. AGU, Fall Meeting, San Francisco, Calif., USA.

Klohn, W. (1979): Selection of appropriate systems for tele-transmission of hydrological data in developing countries. ORSTOM Seminar on satellite teletransmission of hydrological data applied to developing countries' requirements. Sophia Antipolis, France.

Langbein, W. (1972): Evaluation of a stream gauging programme in Kansas. In: Casebook on hydrological network design practice, WMO Publ. No. 324, WMO, Geneva.

Liebscher, H.J. (1982): Hydrological forecasts within Regional Association VI (Europe) of WMO. A report to the WMO, Geneva.

Livanova, N.A., Capotova, N.I., Kondratiev, S.A., Tchecheva, I.L. (1984): Use of Radar Information on Rainfall to Forecast a Hydrograph of runoff (In Russian) Meteorologia i Gidrologia, No. 11, Leningrad.

Miller, J.B. (1982): Requirements for selection of a microcomputer for a HFS. Personal communication.

Moore, R.J. and O'Connel, E. (1978): Real-time forecasting of flood events using a transfer function noise model. Report of Institute of Hydrology, Wallingford, UK.

Miquel, M. (1984): "Les prévisions de crues: methodes, calage, traitement des erreurs", Ecole Nationale des Ponts et Chaussées, Paris.

Natale, L. and Todini, E. (1977): A constrained parameter estimation technique for linear models in hydrology. In: Mathematical models for surface water hydrology, John Wiley & Sons, Chichester, UK.

Nelson, M.L. (1975): Data collection system geostationary operational environmental satellite. Preliminary report, US Dept., of Commerce, NOAA Technical Memorandum NESS 67, Washington, D.C.

Nemec, J. (1972): Engineering Hydrology, McGraw Hill, London.

Nemec, J. and Schaake, J. (1982): Sensitivity of water-resource systems to climate variation, Hydr. Sci. Bul. 27, 3, 9.

Paulson, R. (1978): Review of USGS satellite data-collection research program. American Society of Civil Engineers National Convention, Pittsburgh, Pa.

Pesant, C. (1979): Experience already acquired and immediate projects. ORSTOM Seminar on satellite teletransmission of hydrological data applied to developing countries' requirements. Sophia Antipolis, France.

Sittner, W.T. (1982): Flood forecasting - an introduction. Proceedings, Regional Training Seminar on Flood Forecasting, Bangkok, WMO, Geneva.

Sittner, W.T., Schauss, C.E. and Monro, J.C. (1969): Continuous hydrograph synthesis with an API-type hydrologic model. Water Resources Research, Vol. 5, No. 5, pp. 1007-1022.

Sugawara, M. (1974): On natural disasters - Some thoughts of a Japanese, unpublished manuscript.

Sugawara, M. (1981): How many rainfall stations are needed to measure rainfall for modelling basins in tropical regions? Dic. 2, Proceedings Roving Seminar on Mathematical Models, WMO, Geneva.

Todini, E. and Wallis, J.R. (1977): Using CLS for daily or longer period rainfall-runoff modelling. In: Mathematical models for surface water hydrology, John Wiley & Sons, Chichester, UK.

UNDRO (Office of the UN Disaster Relief Co-ordinator) 1976: Disaster prevention and mitigation. A Compendium of Current Knowledge, Hydrological Aspects, Volume 2, UNDRO/22/76.

US Dept. of Commerce (1976): NOAA National Weather Service, Meteor-burst communications system - Alaska winter field test programme. NOAA TM NWS HYDRO-30, Silver Spring, MD.

US Dept. of Commerce (1976): NOAA National Environmental Satellite Service, User interface manual - GOES data collection system, Washington, D.C.

Water Survey of Canada (1979): Telemetry of hydrometric data, A-Base Review Report No. 47.

Wilson, J.W. and Brandes, A.E. (1979): "Radar measurement of rainfall - a summary", Bull., Ann. Met. Soc. vol. 60, No. 9, September.

WMO (1972): Casebook on hydrological network design practice, WMO Publication No. 324, Geneva.

WMO (1973): Annotated bibliography on precipitation measurement instruments, Report No. 17 on WMO/IHP Projects, edited by J. Rodda, WMO, Geneva.

WMO (1974): Manual on codes, Vol. I - International codes. WMO Publ. No. 306, WMO, Geneva.

WMO (1975): Information on meteorological satellite programmes operated by Members and organizations, WWW Global Observing Systems, Satellite Subsystems. WMO Publ. No. 411, WMO, Geneva.

WMO (1975): Intercomparison of conceptual models used for hydrological forecasting, Operational Hydrology Report No. 7, WMO, Geneva.

WMO (1976): Hydrological network design and information transfer. Proc. int'l seminar, OHR No. 8, WMO Publ. 433, WMO, Geneva.

WMO (1979): Technical Regulations, Volume III, Hydrology, WMO-No. 49, Geneva, Switzerland.

WMO (1980): WMO Technical Regulations - Hydrology and international hydrological codes, WMO Publication No. 7, WMO, Geneva.

WMO (1981): HOMS Reference Manual, internal publication of WMO, Geneva, Switzerland..

WMO (1983): Guide to Hydrological Practices, WMO Publication No. 168, 4th Ed., WMO, Geneva.

WMO (1985): "Activities under the Hydrological Component of Topex", TCP Report No. 20, WMO, Geneva.

WMO (1986): Results of an intercomparison of models of snowmelt runoff. Report to the IAHS Scientific Assembly, Budapest.

WMO (1986) Intercomparison of models of snowmelt runoff. WMO Operational Hydrology Report NO. 23, WMO Publ. No. 646, WMO, Geneva (in press).

Wood, E.F. and O'Connel, P.E. (1985): "Real-time forecasting" in "Hydrological Forecasting", Edited by M.G. Anderson and T.P. Burt, John Wiley, Chichester, UK.

Yoshino, F., Kozeki, D. (1985): Study on short-term forecasting of rainfall using radar raingauge; Hydrology Division, Public Works Research Institute, Ministry of Construction, Tokyo.

Selected list of HOMS components and sequences related to hydrological forecasting

SECTION C, INSTRUMENTS AND EQUIPMENT

C30 PRECIPITATION, RECORDING AND TELEMETERING GAUGES

C30.2.01 Rainfall alarm
C30.3.01 ALERT rain gauge

C33 PRECIPITATION, MEASUREMENT BY RADAR

C33.3.01 Radar rain gauge system

C71 WATER LEVEL OR STAGE

C71.3.01 ALERT river level (float-type) station

SECTION D, REMOTE SENSING

D00.0.07 Snow cover from multi-channel satellite data

SECTION E, METHODS OF OBSERVATIONS

E25 METEOROLOGICAL OBSERVATIONS FOR HYDROLOGY

E25.3.01 The hydrological radar

SECTION F, DATA TRANSMISSION

F00.1.01 Automatic collection and transmission of hydrological observations
F00.1.06 Flood warning device
F00.2.01 Telephone service water level gauge
F00.2.02 Water level transmitter, Telemark
F00.2.03 Telexdat: automatic data collection unit
F00.2.04 Hydra III: regional telemetry unit
F00.2.05 Hydrometeorological data acquisition system
F00.2.06 Flood warning system
F00.2.07 Automatic data acquisition system
F00.2.08 Data acquisition and data transmission via satellite (the Argos system)
F00.2.09 Satellite data collection platform
F00.2.10 Hydrological data transmission system
F00.2.11 Data transmission via geostationary satellites
F00.2.12 Data transmission via polar orbiting satellites
F00.2.13 Telemetering station land radio link
F00.2.15 Satellite direct readout ground stations

F00.2.16	Snotel data acquisition project
F00.3.01	Data collection and transmission system (Ott Allgomatic)
F00.3.02	River information system
F00.3.03	Level and rain digital telemetering system
F00.3.04	Automatic network for collecting real-time hydrological data
F00.3.05	ALERT data collection substation

SECTION J, HYDROLOGICAL MODELS FOR FORECASTING AND DESIGN

J05	RAINFALL-RUNOFF MODELS FOR GENERAL USE
J05.1.01	Tank model
J05.1.02	Runoff calculation by the storage function method
J05.1.03	Seasonal forecast of inflow to a lake
J05.2.01	Runoff model for cultivated soils
J05.2.02	Streamflow synthesis and reservoir regulation (SSARR)
J05.2.03	Nonlinear cascade hydrological model (NONLIN)
J05.3.01	Manual calibration program (NWSRFS-MCP)
J05.3.02	Sacramento soil moisture accounting model (NWSRFS-SAC-SMA)
J05.3.03	Sacramento model modified for use in the upper nile basin project
J05.3.04	Multipurpose unsteady flow simulation sytem (MUFSYS 3)
J12	RAINFALL-RUNOFF MODELS FOR FORECASTING
J12.1.01	Flood forecasting by graphical correlation
J12.1.02	Deterministic rainfall-runoff model for basins of 100 to 1000 sq.km.
J12.1.04	Flood peak forecasting by a grapho-analytic technique
J12.1.05	A method to forecast the spring flood volume
J12.1.06	Model for operation (MODRIE)
J12.2.01	Water supply streamflow and storage forecasts
J12.2.04	A conceptual watershed model for flood forecasting
J12.2.05	Forecasting inflows to a lake
J12.2.08	Conceptual watershed model (the HBV model)
J12.2.09	Model to forecast rainfall floods
J12.2.10	Model for the calcuation of snowmelt and rainfall runoff
J12.2.11	Method for short-term forecasts of discharges in mountain rivers
J12.3.01	Real time streamflow forecasting model (MISP)
j12.3.02	Application of hydrologic models for river forecasting
J40	PROCESS MODELS
J40.1.01	Snowmelt-runoff model (SRM)
J50	STOCHASTIC MODELS, GENERAL
J50.2.04	CLSX (constrained linear system extended) model

J70 STREAM FLOW ROUTING, GENERAL

J70.3.02 Hydrodynamical river model (DC2)

J72 STREAM FLOW ROUTING, SINGLE CHANNEL

J72.1.01 Channel routing Muskingum-Cunge method
J72.2.01 Lag and K routing (NWSRFS-LAG/K)
J72.3.01 Channel routing - implicit solution of full equations

J74 STREAM FLOW ROUTING, MULTIPLE CHANNEL

J74.2.01 Dynamic wave operational model (DWOPER)
J74.2.03 Method of unsteady flow calculation in braided river beds

HOMS sequences related to hydrological forecasting

007 Dynamic flow forecasting
009 Rainfall/runoff simulation using a soil moisture accounting level
010 ALERT flood warning system
011 Satellite data collection/relay systems

Results of a WMO survey of 17 countries of Europe, covering a total of 74 HFS (Liebscher, 1981)

Of a total of 74 forecasting procedures (models), 35 use statistical or stochastic approaches (either multiple correlation or autoregressive function), 16 use routing procedures, often based on empirical concepts, and 23 use conceptual (deterministic) hydrometeorological approaches.

The basins covered by operational forecasting systems have a size of between 100 and 1,000,000 km^2. The size of 10,000 km^2 is the most frequest (21 cases). Basins of less than 100 km^2 are rarely forecast.

The purpose of the hydrological forecasts is first of all flood protection (57 cases). Energy production is second (25 cases), followed by navigation (16 cases), and water supply and sanitation (16 cases), irrigation (8 cases), water pollution control (5 cases), and ice problems (5 cases).

The most frequent forecast elements are water level (54 cases), then discharge (46 cases), and volume of runoff (27 cases). Forecasting of ice, groundwater level, and water quality is rarer.

Hydrographs of floods are in the form of forecast output in 41 cases, a continuous hydrograph in 32 cases. In 34 cases, the subject of the forecast is the time of occurrence. In 15 cases, minimum flows are forecast.

Most forecasting systems provide short-term forecasts. In 24 cases the forecast is up to 24 hours and in 27 cases the forecast ranges between 1 day and 1 week. Medium-term forecasts (11 cases) and long-term forecasts (10 cases) are surprisingly frequent.

While a forecast is permanent in 42 cases, the forecast is restricted only to flood season in 20 cases, and it operates only after exceeding or - in the case of low-water forecast - falling below certain threshold values of discharge in 11 cases.

Computers or programmable desk calculators are used in most of the forecasts (45 cases). Graphs (6 cases) and manual and tabular computation (20 cases) are used in older systems.

The most frequent input variable for forecasting is precipitation (53 cases); snow-water equivalent is used as an input parameter in 29 cases, followed by data on air temperature (19 cases), soil moisture (18 cases), evaporation (10 cases), radiation (9 cases) and other meteorological elements (10 cases). The use of groundwater levels, water-quality data and water temperature as input for forecasting is rare.

The use of forecast data as input is surprisingly high. Precipitation forecasts are used in 39 cases, discharges in 35 cases, water levels in 22 cases, and air temperature in 14 cases.

Data collection is done manually in 47 cases. Automatic stations are in operation in 38 cases. In many forecasting systems data collection is done manually as well as by automatic stations. In 23 cases data are recorded. Modern techniques such as radar (9 cases) or remote sensing techniques by aircraft (2 cases) or by satellites (1 case) are not frequently used.

In 68 cases the transmission of data is by telephone. Mail is used in 12 cases. Data are transmitted by radio in 20 cases and by telex in 22 cases. Satellites are used only in 3 cases.

Example of a HFS in the Pampanga River Basin, Philippines

1. COUNTRY: Philippines

2. RIVER BASIN

2.1 Basin name: Pampanga River basin

2.2 Total area: 10540 km^2 of which 8550 km^2 is the basin area draining to the lowest forecast point at Sulipan - Apalit.

2.3 Index map: - A map (1:1,000,000) showing the location of the basin and the main river is given in Appendix 1.

3. HFS

3.1 Map: A map (1:250,000) showing the system and data transmission network is given in Appendix 2.

3.2 General information

(a) *Purpose of the system*

The Pampanga River is one of the two major river systems in the Philippines. The area suffers considerable flood damage almost every year; about one fourth of the total flood damage in the Philippines may be attributed for the Pampanga River. The basin is densely populated with various important development projects. The purpose of the system is to provide flood forecasts with a view to reducing flood damage.

Target areas to be protected by the system are places surrounding the Candaba swamp area, lower delta areas and places on the western side of the levee separating the swamp.

(b) *History of the system*

In 1966/1967, a WMO/ESCAP survey mission recommended the development of a pilot flood forecasting and warning system in the Pampanga River basin. This recommendation was subsequently endorsed by the Typhoon Committee in 1968. The Government of the Philippines accordingly decided to select the Pampanga River basin for pilot flood forecasting in view of its importance to the economy of the country.

Within the framework of bilateral co-operation, a team of Japanese experts carried out a feasibility survey and submitted a detailed report on the project in 1970. The system was subsequently established and began operations in September 1973.

(c) *Resources of the forecasting centre*

Organization

The system is operated by the National Flood Forecasting Office of the Philippine Atmospheric, Geophysical and Astronomical Services Administration (PAGASA). The number of operational employees include: seven hydrologists, five telecommunication experts and three support staff.

Computer facilities

A Hewlett-Packard desk computer is used for operational purposes. An 1130 IBM computer and HP computer are being used for developmental research.

Operating cost

The average annual budget in recent years amounts to about US$10,000.

Capital cost

The capital cost of equipment, housing facilities, training costs, etc., is estimated at US$688,000 at 1981 prices.

Installation cost

The installation cost is estimated at US$65,000 at 1981 prices.

3.3 Data collection system

(a) *Rainfall stations*

Twenty-three ordinary stations; 27 self-recording stations and 14 self-recording and telemetering stations.

(b) *Water-level stations*

Ten stations with staff gauges and 8 stations with water level recorders.

(c) *Streamflow stations*

Discharges are measured:

(i) daily at 5 stations (all having water-level recorders)

(ii) twice daily at 8 stations (7 stations with water-level recorders and one station with only a staff gauge)

(iii) three times a day at 4 stations (3 stations with water-level recorders and one station with a staff gauge).

During floods, the frequency of observations range from hourly to 6-hourly. Data has been collected, but with intermittent disruptions since the establishment of the system in 1973.

3.4 Data transmission system

The telemetering system and data transmission system is given in Appendix 2. A 150 MHz band radio wave frequency is used.

3.5 Data processing system

A subjective method is used for checking the incoming data as input to the forecasting model.

3.6 Receipt and use of meteorological forecasts

Forecasts of tracks/intensity of tropical cyclones, generalized forecasts of rainfall amounts, durations and types of other meteorological information including satellite and radar data are received and used in determining a subjective precipitation forecast over the basin, which forecast is then used as input to the forecasting model.

3.7 Flood-forecasting model

The forecasting model used is the Tank model (two tanks in series) and Empirical Water Stage Correlation. The Tank model is used to compute forecast discharges at Sulipan-Apalit with the catchment area of the Pampanga River above the station represented by a combination of two tanks in series. See list of references for more detailed information on the Tank model.

3.8 Warning system

The system envisages forecasting the flood level at vulnerable places in the lower reaches of the river one day in advance, and if possible two days. Forecasts of water stages are prepared with reference to Arayat, Candaba and Sulipan Stations (Appendix 2). Forecast values are then interpolated with reference to the target areas.

Flood warnings are normally released twice a day (9.00 a.m. and 9.00 p.m.) and are relayed promptly by telephone to the Office of Civil Defense (OCD), Ministry of National Defense and Ministry of Public Works (MPW) in Manila. Warnings are also broadcast over a government-operated station with a terminal at PAGASA. The warnings are also furnished to the news media.

3.9 Forecast users

The users of the forecasts and the purpose(s) for which they use the forecasts are listed below:

(i) Office of Civil Defense (OCD) - Flood warning dissemination to other agencies concerned or to general public and use of the warnings for managing disaster operations.

(ii) Ministry of Social Services and Development (MSSD) and Philippine National Red Cross (PNRC) - Execute plans in accordance with the stages of preparedness and to provide relief assistance.

(iii) Ministry of Public Information (MPI) - Asked to provide public information or warning dissemination.

(iv) Ministry of Public Works (MPW) - Flood monitoring for flood fighting.

(v) Private agencies/individuals - For protection of life and property.

(vi) New media.

3.10 Problems with the system

The main problems are the unavailability of necessary spare parts and insufficient local expertise for the maintenance of the system. Plans are being made for the development and improvement of the flood forecasting model/procedures.

4. LIST OF REFERENCES

Sugawara, M. (1961) - On the analysis of runoff structure about several Japanese rivers. Jap. J. Geophysics, Vol. 2, No. 4.

Sugawara, M. et al (1974) - Tank model and its application to Bird Creek, Wollombi Brook, Bikin River, Kitsu River, Sanaga River and Nam Mune. Research Notes of the National Research Center for Disaster Prevention.

Sugawara, M. et al (1975) - Runoff analysis of snowy mountainous regions in Japan. IBID.

Sugawara, M. (1979) - Automatic calibration of the Tank model.

Hydrological Science Bulletin, No. 24.3.

ANNEX 3 APPENDIX 1

Location of Pampanga River Basin

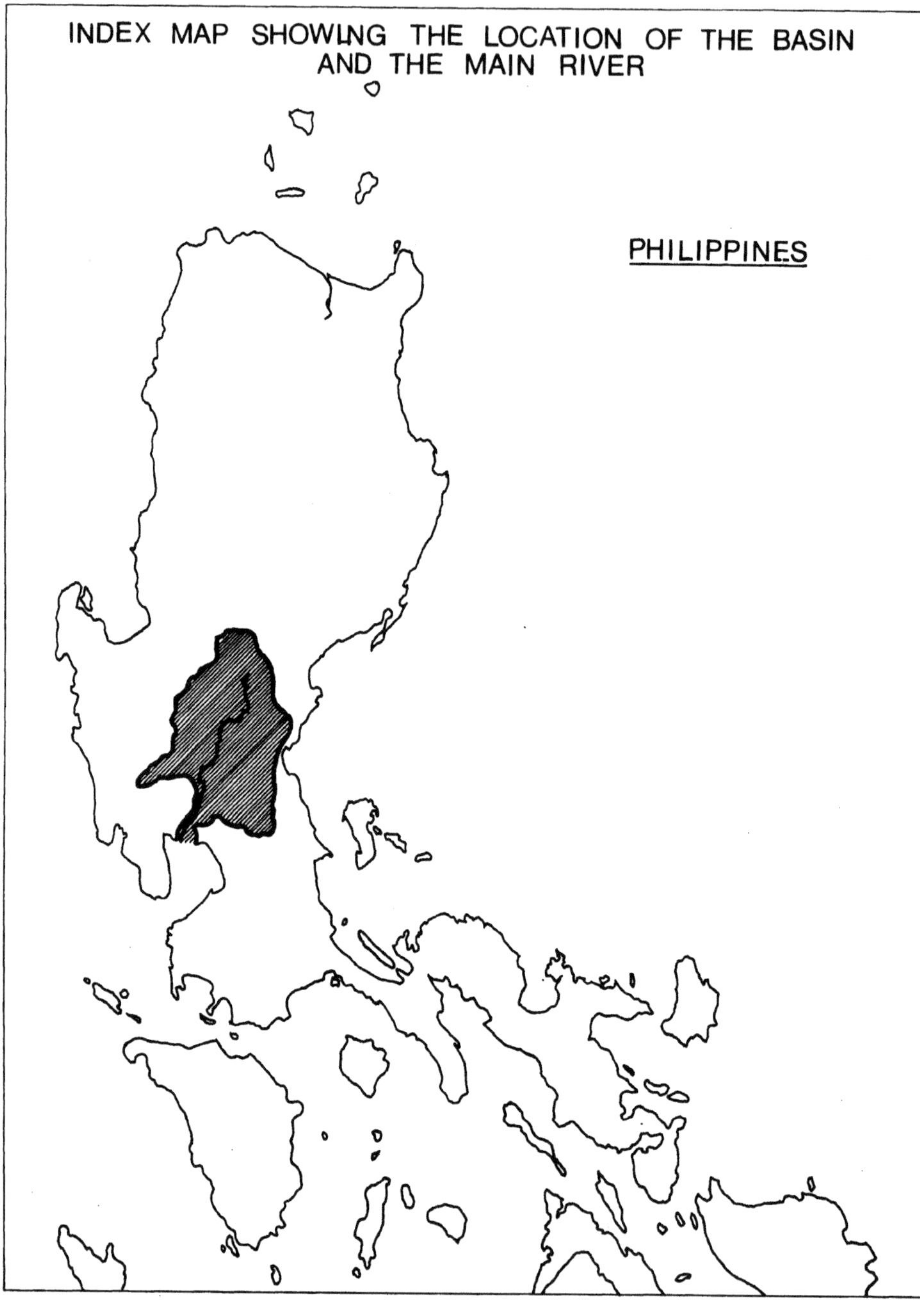

ANNEX 3 APPENDIX 2

Telemetry system.

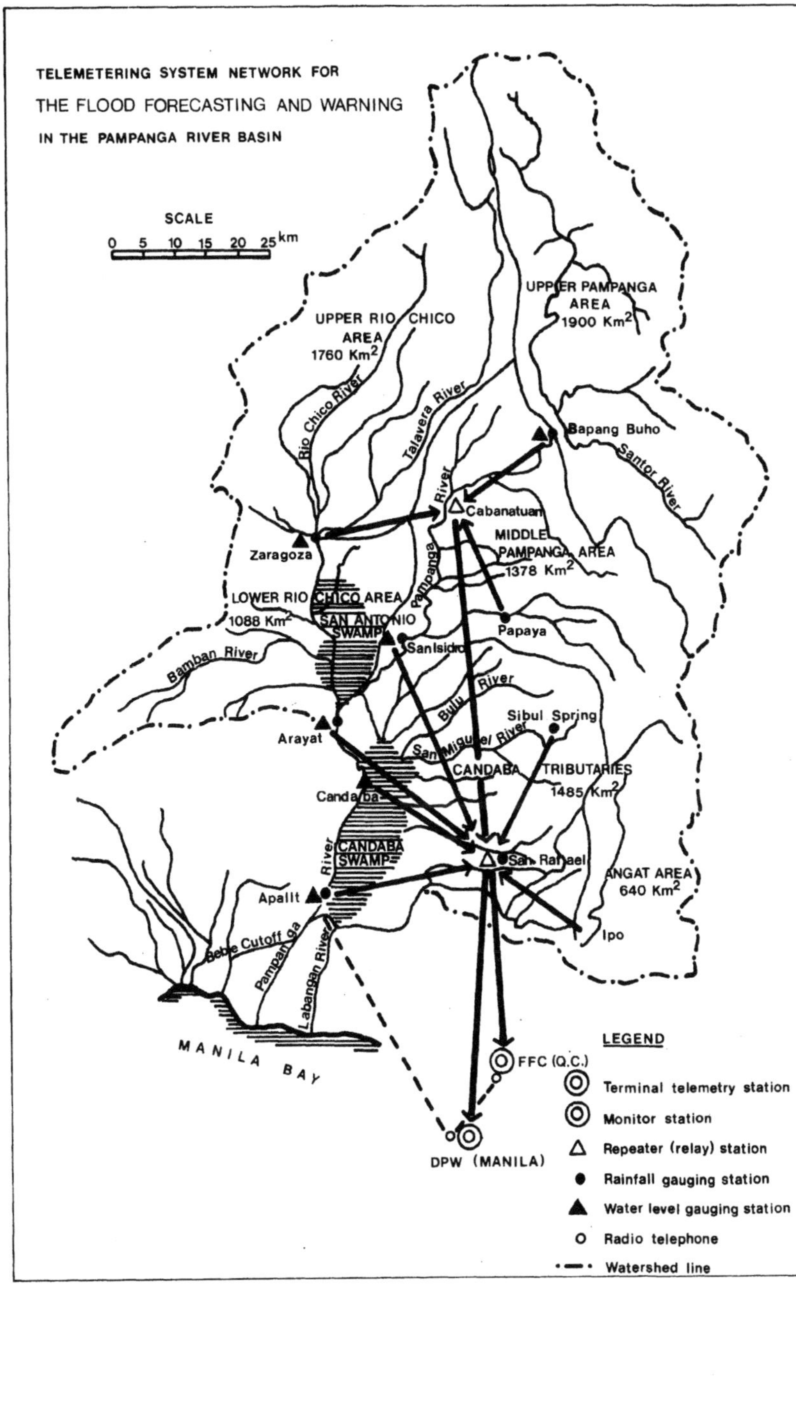

Example of a HFS in the Indus River Basin, Pakistan

1. COUNTRY: Pakistan

2. RIVER BASIN

2.1 Basin name: Indus River Basin

2.2 Total area: 970,000 km^2 of which 469,000 km^2 in Pakistan

2.3 Index map - A map showing the location of the basin and the main river is given in Appendix 1.

3. HFS

3.1 Map - A location map of an improved hydromet network is given in Appendix 1.

3.2 General information:

(a) *Purpose of the system*

Pakistan is almost totally economically dependent upon its water resources as represented by the Indus River Basin and its five rivers: Indus, Jhelum, Chenaab, Ravi and Sutlej. However, these rivers have also produced devastating floods which, since Independence (1947), have been of record or near record magnitude. The system is intended to:

(i) optimize planning, operation, regulation and management of the river system (Mangla, Chasma and Tarbela reservoirs and linked irrigation systems);

(ii) minimize the damages to the vital hydraulic structures and irrigation works;

(iii) provide safety for people and livestock against floods; and

(iv) aid in planning and taking measures to prevent damage to property.

(b) *History of the system*

Prior to 1972, there were many stream gauging and climatological stations being operated by different agencies. However, after the catastrophic floods in August 1972, it was concluded that the system then in existence needed a complete re-appraisal. This conclusion resulted in two World Meteorological Organization (WMO) representatives visiting Pakistan in March 1974 to assess the situation. Then in 1975, the WMO/UNDP joint co-operative projects (PAK/74/027 and PAK/75/WMO/FIT, "Improvement of the River Forecasting and Flood Warning System for the Indus River Basin in Pakistan") were initiated. The three components of these projects are: Component I - expansion and modernization of the basic Hydrological Data Collection and Processing System; Component II - establishment of a radar at Sialkot and a meteorological satellite readout station at Lahore; Component III - development of river flow and flood-forecasting techniques. Originally a snowmelt hydrology component was included, but was dropped early on with the intention of taking it up again later. The three components were completed in October 1980, after which the system was taken over by the Government of Pakistan. The system was test run during the 1980 flood season. A second phase aiming at an improvement of the system and seasonal forecasts from snowmelt was started in 1985.

(c) *Resources of the forecasting system*

Organization

The two projects' activities were carried out as a single project. WMO was designated as the executing agency with two principal counterpart agencies: Pakistan Meteorological Department (PMD) and the Water and Power Development Authority (WAPDA). A technical committee was formed in 1976 to co-ordinate the various agencies. They have continued to execute this responsibility after the completion of the project while the responsibilities of maintaining and inspecting the system were taken over by the Government of Pakistan in 1980.

Computer facilities

Two "Nova Computer - 3/12s" were installed, one in Lahore and one in Sialkot. A "Silent 700" serves as the interface medium with the Automatic Data Telemetering. Also, an Automatic Picture Transmission ("Alden APT-38 type") was installed in 1978 in Lahore. There is also a 9825A Hewlett-Packard calculator. In the second phase three microcomputer IBM-XT have been installed.

Operating cost

The 14 international employees (three team leaders, four experts and five consultants) had 21 national counterparts who included 11 meteorologists (hydrological forecasters), five hydrological engineers, plus support staff.

Capital cost

The total budget contributed by the Dutch Government and UNDP was US$2,503,297. In addition, the Government of Pakistan contributed Rs. 3,813,000.

3.3 Data collection system

This is continuously running, self-reporting system based on discharge input data. There is one weather radar for the upper catchment rivers monitored as hourly rainfall amounts. There are 30 forecast sites which rely on river level sensors. The accumulators for these report on six cm increments (sensitivity - 3, 6 12 and 24 cm increments). Transmitters report every one-and-a-half hours.

The telecom system has 36 stations with VHF automatic data transmission (observation times range from continuously to six hourly); seven stations with HF voice communications and eight stations without telecommunication facilities.

The Automatic Picture Transmission Image has been operative since August 1978. It relies on two satellites (TIROS-N and NOAA-6) which pass twice over a point in 24 hours.

3.4 Data-transmission system

The data transmission system is given in Appendix 2.

3.5 Data-processing system

Data processing is incorporated in the forecasting software.

3.6 Receipt and use of meteorological forecasts

The weather radar at Sialkot covers the upper catchments of the Chenab, Ravi, Indus and Sutlej Rivers. Hourly rainfall amounts are supplied to FF and WC for 17 sub-catchments. Radar and satellite data are used for making daily forecasts.

3.7 Flood-forecasting model

The facilities provide real-time data collection and transmission to FFWC at Lahore as inputs to the computer-based flood-forecasting and routing models. These models are of two types: (1) the Rainfall-Runoff Model which is used for the catchment areas above the rim stations; and (2) the Flood Routing Model which is used to calculate the movement and information on flood waves from the rim stations to the sea.

Both Constrained Linear System (CLS) and Antecedent Precipitation Index Continuous (APIC) models have been calibrated for forecasting the discharge at rim stations for the Indus, Jhelum, Chenab, Ravi and Sutlej Rivers. The lead time of the forecast is one day. The CLS model is basically a flood-routing model which minimizes errors normally associated with the unit hydrograph and determination of areal mean rainfall. It can be used for routing and is suitable for the combination of routing and lateral inflow. The APIC model utilizes an incremental rainfall-runoff relationship comprised of: the rainfall-runoff relationship, the unit hydrograph and the relationship for computing groundwater flow. Thus the relationship is derived by first calibrating the catchment to a conventional total storm relationship and then converting this to and incrmental type.

The flood routing model divides the Indus River Basin into 30 river reaches each with a set of "externals" in the form of inflow related to the tributaries or link canals, etc., outflows of offtakes in the form of canals, etc. The model provides flood routing of the entire basin. The flood forecasts can be plotted for individual reaches. A lead time of 10-15 days is available in forecasting the flood/flows at Guddu based on the observed flood/flows at the rim stations. Thereafter these can be updated on a daily basis. Two types of flood routing models have been calibrated based on lumped or distributed parameters.

3.8 Warning system

With the help of these models, weather radar and telemetering systems for the river flow and flood forecasting are performed for the "rim" stations approximately 24 hours in advance; as well as for 30 sites of the Indus River System for between 12 hours to 12 days ahead, depending upon the location of the site. A forecast lead time of five to seven days can be achieved for the dangerous locations at the confluences. The low incidence rate of river floods during the 1980 monsoon season (only one high flood on the Ravi River) has not permitted a full evaluation of the accuracy of the forecasts.

3.9 Forecast users [see above section 3.2(a)]

3.10 Problems with the system

There is a problem with variability in the equipment and methodologies being utilized for discharge and sediment transport measurements. This significantly reduces the accuracy of forecasts. Three mobile stations have been proposed to increase the reliability of discharge sediment curves.

There is also a need for a well-trained and experienced hydrologist and hydrometeorologist to operate and improve the telemetering network.

4. LIST OF REFERENCES

(a) Water and Power Development Authority/Pakistan, Improvement of River Forecasting and Flood Warning System for the Indus River Basin in Pakistan, Brochure of the Hydrology and Investigation Directorate, WAPDA, Lahore, Pakistan, 1980.

(b) WMO/UNDP, Improvement of the River Forecasting and Flood Warning for the Indus River Basin in Pakistan: Project Findings and Recommendations, PAK/74/027 and PAK/75/WMO TF, WMO/UNDP, Geneva 1980.

(c) Government of Pakistan, Project Proposal - Component I, Revised and Approved 7 May 1977.

ANNEX 4 APPENDIX 1

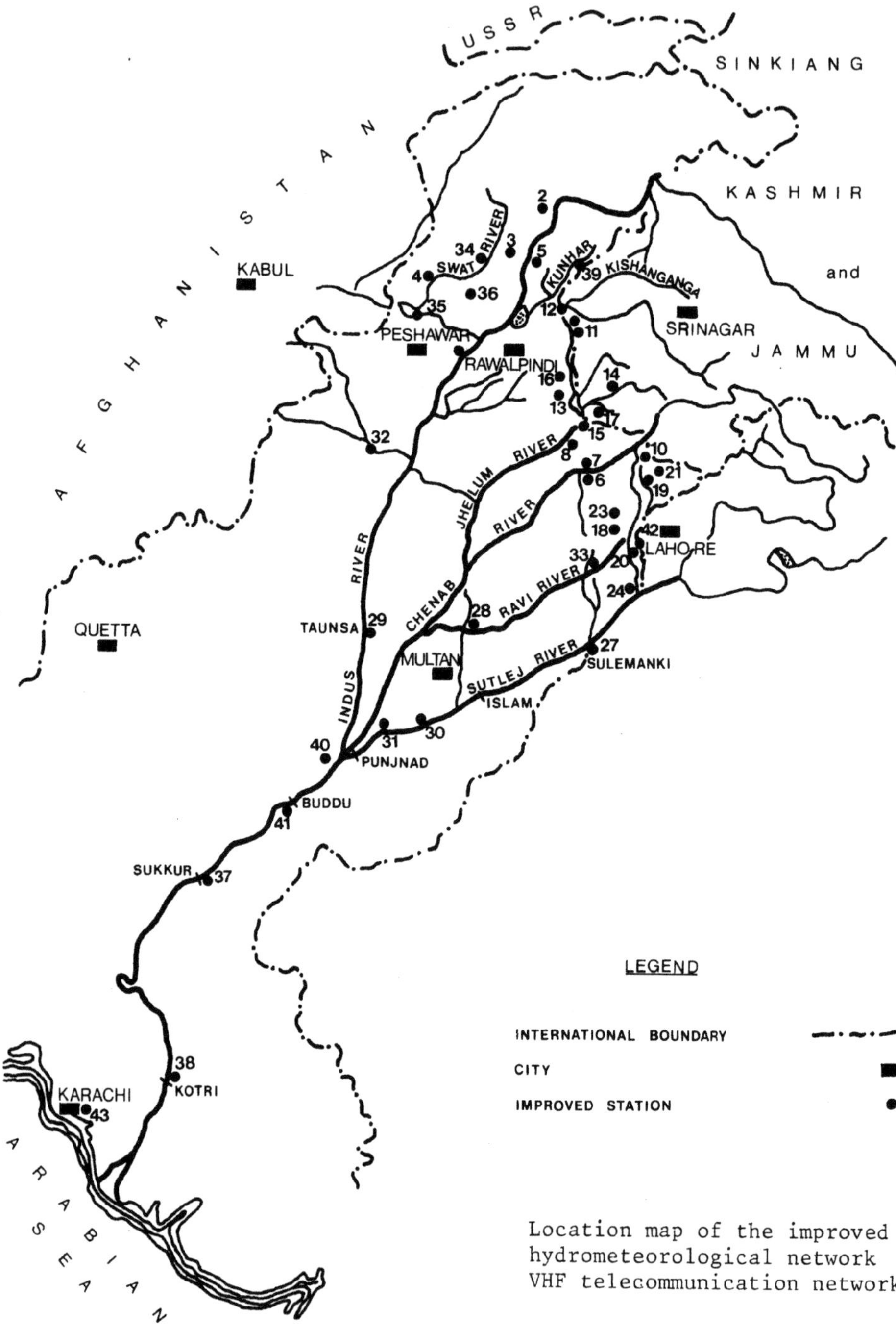

Location map of the improved hydrometeorological network VHF telecommunication network

ANNEX 4 APPENDIX 2

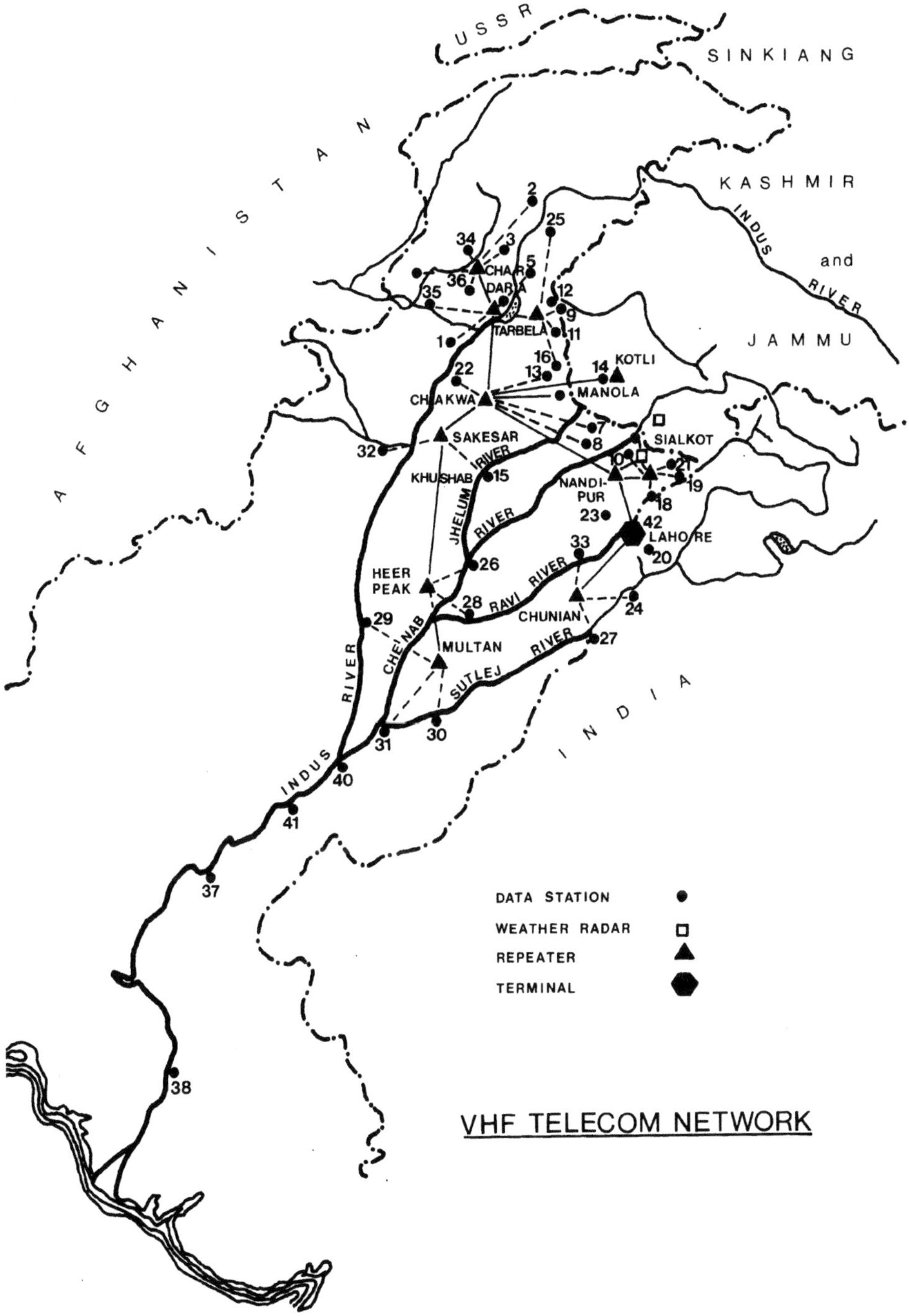

VHF TELECOM NETWORK

Examples of HFS in the Amazon River Basin, Brazil - Lower and Middle Amazon and the Tocantins

1. COUNTRY: Brazil

2. RIVER BASIN

2.1 Basin name: Lower and middle Amazon

2.2 Total area: 1,980,000 km^2

2.3 Index map: A map showing the total basin and the area of the model is attached as Appendix 1.

3. HFS

3.1 Map - Figure 21 shows a map of the division of the area used for applying partial models. Appendix 2 shows a map with the observing-stations network of one of these sub-basins.

3.2 General information

(a) *Purpose of the system*

The Amazon river is the largest in the world and its basin extends to seven countries. The middle and lower Amazon has an extremely low slope. For example, the Solimoes (upper Amazon), as it enters Brazil, has an elevation of 62 meters above sea level, which results in a medium slope of 2×10^{-5}.

As a consequence of this, over its whole Brazilian sector, and in the delta in particular, the river forms large islands and extensive flood plains (varzeas). These "varzeas" are practically the only zones with good agricultural potential in the lower and medium Amazon and are subject to periodic flooding. Most of the cities in this zone, and particularly Santarem, Parintins and Obidos, are partially located in these flood plains.

Forecasts made with ample lag (up to one month) are therefore required for planning the agricultural production and cattle raising in the "varzeas" as well as for civil defense of most of the cities in the area. Low water forecasts are also required for river navigation purposes, since water courses are the most important means of transportation in the area.

(b) *History of the system*

Inception of the WMO/UNDP project "Hydrology and Climatology of the Brasilian Amazon River Basin" started in 1972, but actual field operation only in 1977. One of the objectives of the projects was the development of this forecasting system. Forecasts started in 1979 using a multiple-correlation model and transmission of the data via telephone or through the meteorological radiotelecommunication (SSB) of INEMET (National Meteorological Institute). The development of a model was sub-contracted by WMO to the Argentinian firm EGASAT. The model was completed in 1982 and intalled in the computers of SUDAM (Superintendency for the Development of the Amazon) in Belem. The model started to be used operationally in 1983. Plans for using the GOES satellite as a relay station for the main communications link had to be cancelled due to lack of funds, although such a system was developed and is operational for the Tocantins river basin (adjacent to the Amazon basin).

(c) *Resources of the forecasting centre*

Organization

The system is operated by the "Projeto de Hidrologia e Climatologia" of SUDAM, at Belem, Brazil. The number of operational employees include: three hydrologists, three computer technicians and six hydrological technicians.

Computer facilities

A Hewlett-Packard 9845B desk-top computer is used for operating the PLUMUS model (see 3.7) and the hydrodynamic model operates in a Burroughs 2700 computer.

Operating cost

No figures available.

Capital cost

The capital cost of the HP computer and the development of the model is estimated at $120,000. This does not include the observing network which was available prior to the installation of the system.

Installation cost

See above.

3.3 Data-collection system

(a) Rainfall stations: 96 stations (13 with recorders)

(b) Climatological stations (for computing potential evapotranspiration): 13 stations

(c) Water-level stations: 29 stations. Twenty-seven of these take stage/discharge measurements. The level is read once or twice a day, which is sufficient taking into account the size of the basin.

3.4 Data-transmission system

Telephones or telegraph to two main collecting bases: Manaus and Belem; teletype link between Manaus and Belem.

3.5 Data-processing system

All data are stored daily using flexible disks of the Hewlett-Packard 9845 system. Two mass-memories are available, so all data are recorded twice.

3.6 Receipt and use of meteorological forecasts

Alerts of the possibilities of heavy precipitation on the basin are received from the Brazilian Meteorological Institute (INEMET) in Brasilia, via teletype.

3.7 Flood-forecasting model

A relatively complex model has been developed consisting of two stages:

(a) For each of the sub-basins indicated in Figure 21, the flows are calculated with the "PLUMUS" model. PLUMUS consists of two sub-models: BALAN, which is a rainfall-runoff storage model and MUSK for flow routing which is basically a Muskingum routing procedure.

(b) A hydrodynamic model is then used to route the flow between Manaus and Santarem (Appendix 4). The model consists fundamentally of the solution to the St. Venant equations (Appendix 3).

3.8 Warning system

Forecasts covering up to 21 days are issued by the forecasting office every seven days; these are transmitted by telex or telegram by SUDAM to the authorities of the cities of Manaus, Parintins and Obidos as well as to the Civil Defense Authorities in Manaus and Belem. It is planned in the future to include other users.

3.9 Forecast users

(a) SUDAM, Belem and Manaus

(b) Civil Defense organizations

(c) Municipalities of the area

(d) Brazilian Navy (River Navigation)

(e) Eletronorte (Power company responsible for electricity production).

3.10 Problems with the system

The two main problems are:

(a) The communication system is not automated, therefore data from only a part of the total network is received on a regular basis; and

(b) the complexity of the model requires a much larger computer than the HP 9845 or the Burroughs 2700 in order to reduce the time of computation presently required (several hours).

4. LIST OF REFERENCES

WMO (1982) - Mathematical Models of the Hydrological Regimes of the Medium and Lower Amazon. EGASAT Final Report for Project BRA/82/031.

WMO (1982) - Mathematical Models of the Hydrological Regimes of the Medium and Lower Amazon. Manual of Operation, Project BRA/82/031.

1. COUNTRY. Brazil

2. RIVER BASIN

2.1 Basin name: Tocantins River

2.2 Total area: Approximately 700.000 km^2

2.3 Index map - A map showing the basin is attached as Appendix 6.

3. HFS

3.1 Map - The map in Appendix 6 shows the forecasting points and the location of the telemetering stations.

3.2 General information

(a) *Purpose of the system*

The Brazilian Government through Eletronorte is constructing at Tucurui (see Appendix 4) the third largest hydroelectric power plant in the world (6000 MW). Forecasts are needed during the construction phase and later for the operation of the dam. The forecasting sytem will serve both purposes.

The lower Tocantins and its main tributary the Araguaia have very low slopes and extensive flood plains in which several cities are located. The forecasting system will also serve to give appropriate warnings to these cities, and in particular to Marabé (see map) for civil protection purposes.

(b) *History of the system*

Forecasts using a simplified (only routing component) SSARR (US Corps of Engineers) model were started by Eletronorte in 1977. In parallel, forecasts using simple statistical models (multiple correlation type) were started in 1978 by WMO/UNDP project BRA/72/010 "Hydrology and Climatology of the Brazilian Amazon River Basin" (PHCA). The Tocantins River is a tributary to the Amazon delta.

In July 1979 an agreement was reached between PHCA and Eletronorte to co-ordiante efforts. During one year both models were used for forecasts with 1, 4 and 7 days' advance. The results obtained with the statistical model were far better than those obtained with the simplified SSARR as shown in tables of Appendix 7.

Up to this stage the data of only four upstream stations were used (shown in Appendix 4) and telephone or SSB radio transmission was used, which resulted in frequent telecommunication failures. In view of this a study was made under PHCA to select a better telecommunications system. The following systems were considered:

(i) Conventional radio;
(ii) Microwaves;
(iii) Polar-orbiting satellites;
(iv) Geostationary satellites;
(v) Meteorburst.

The GOES (US Geostationary Operational Environmental Satellite) was selected. Ten data transmitting platforms (DCPs), each with five sensors (water level, rainfall, pressure, temperature and humidity) were installed in the locations shown in Appendix 8. The system became operational in 1982 and was immediately used for forecasting purposes. The receiving station is located at Sao Paulo from where the data are transmitted via teletype to Eletronorte in Brasilia and to PHCA at Belem.

(c) *Resources of the forecasting centre*

Organization

The system is operated by the "Projeto de Hidrologia e Climatologia" of SUDAM, at Belem, Brazil. The number of operational employees include: three hydrologists, three computer technicians and six hydrological technicians.

Computer facilities

Two Hewlett-Packard desk-top computers are available: one HP 9831, with 24 K of memory, and one HP 9845, with 126 K of memory.

Operating cost

No figures available.

Capital cost

The capital cost of the HP computers is estimated at $40,000. The cost of the telemetry station and the receiving stations was of the order of $200,000 ($12,000 per station plus $80,000 for the receiving stations). This does not include the cost of the civil engineering works which were borne by Eletronorte.

Installation cost

The cost of the supervision of the installation was of the order of \$30,000. No figures are available for the local costs, which were considerably higher.

Data-collection system

Ten telemetering stations with five sensors (water level, rainfall temperature, air humidity and pressure) (see above) and one receiving station.

Data-transmission system

The GOES satellite was used for relaying the messages to Sao José dos Campos (near Sao Paulo). Teletype is then used for transmission to the forecasting centre in Brasilia and Belem and from there to the final users.

Data-processing system

Data are stored daily in cassettes, using the keyboard of the HP 9831. Back-up memory is provided by two flexible disk memories.

Receipt and use of meteorological forecasts

Alerts of the possibilities of heavy precipitation on the basin are received from the Brazilian Meteorological Institute (INEMET) in Brasilia, ia teletype.

Flood-forecasting model

A simple forecasting model of the type:

$$Y = a_o + a_1X_1 + a_2X_2 + a_3X_3 + a_4X_4$$

is used, where:

Y = Discharge at the forecasting point at day t $+\Delta$ less discharge at the forecasting point at day t; t being the data on which the forecast is made and the period of the forecast (Δ= 1, 4 and 7 days)

X_1, X_2, X_3, X_4 = Discharges at sites 1, 2, 3, 4 at day t $+\Delta$ less discharge in the same sites on day t

a_0, a_1, a_2, a_3, a_4 = Numerical coefficients obtained by multiple regression

The availability of ten telemetering stations now makes it possible to apply more advanced models; two of which are being developed: (a) the PLUMUS model used in the middle and lower Amazon; and (b) the complete SSARR, with its rainfall-runoff component (formerly only the routing part had been used).

Warning system

Forecasts for one, four and seven days are issued on all working days (everyday during floods). These are transmitted from the PHCA by telex to the Tucurui dam for the construction management and to SUDAM at Maraba for civil defense purposes.

Forecast users

Eletronorte, construction of Tucurui dam, Sudam, civil defense of Maraba and other towns.

Problems with the system

The system is now operating without major problems.

4. LIST OF REFERENCES

PHCA(1978) - Aplicaçao de Modelos Estatisticos para à previsao de niveis na Bacia Amazonica (Application of statistical models for level forecasts in the Amazon Basin), Publication No. 4, Belem, Brazil.

PHCA (1980) - Aplicaçao de Modelos Estatisticos à Bacia do Rio Tocantins (Application of statistical models to the Tocantins River Basin), Publication No. 28, Belem, Brazil.

UNDP/WMO (1982) - Implementation of GOES Telemetry in the Tocantins Basin, Technical Report by Robert A. Halliday, WMO, Geneva.

Basso, E. (1982) - Hydrology and Climatology of the Amazon, Water World, New Delhi, Vol. IV, No. 3, pp. 5-12.

ANNEX 5 APPENDIX 1

Amazon River map and river system.

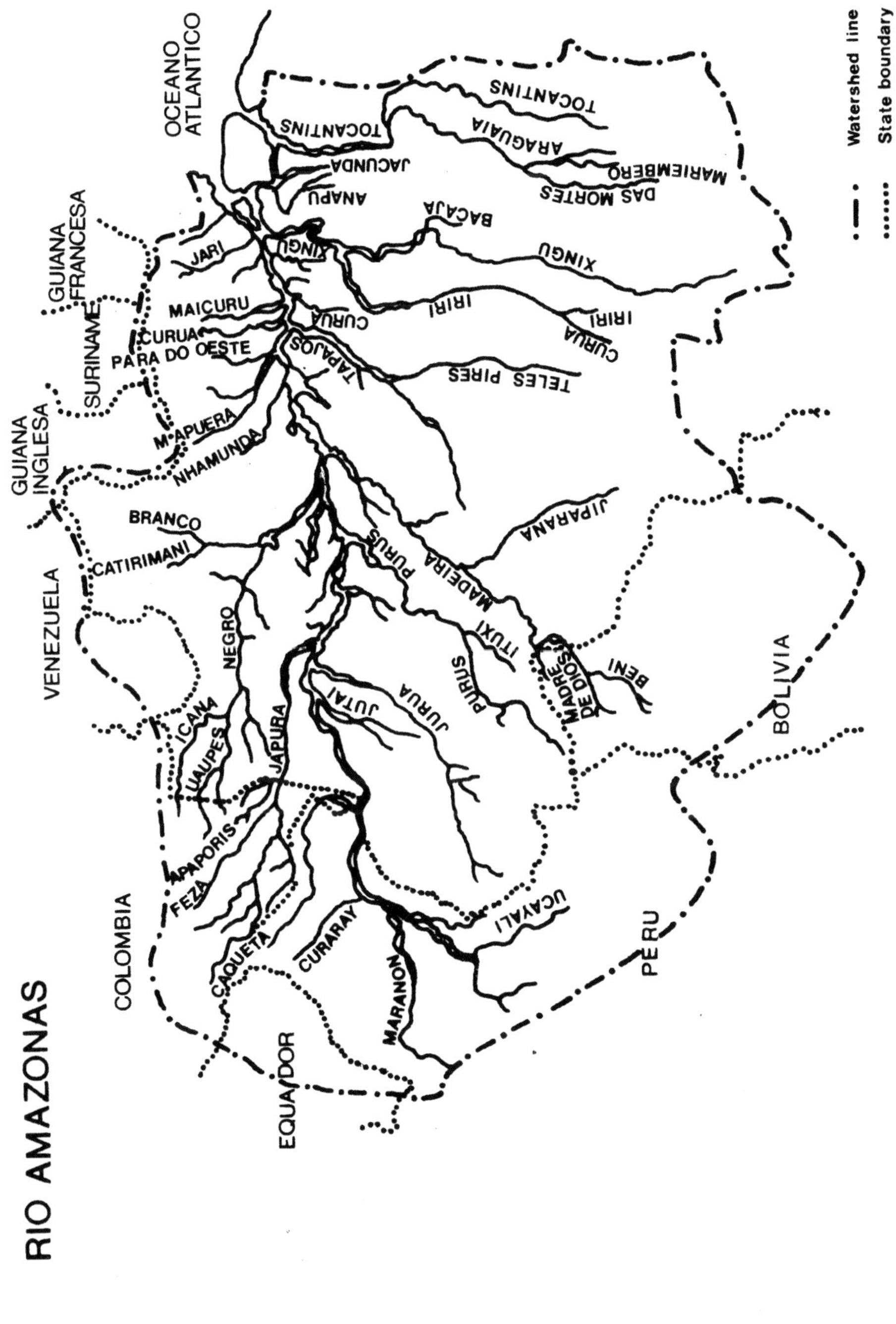

ANNEX 5 APPENDIX 2

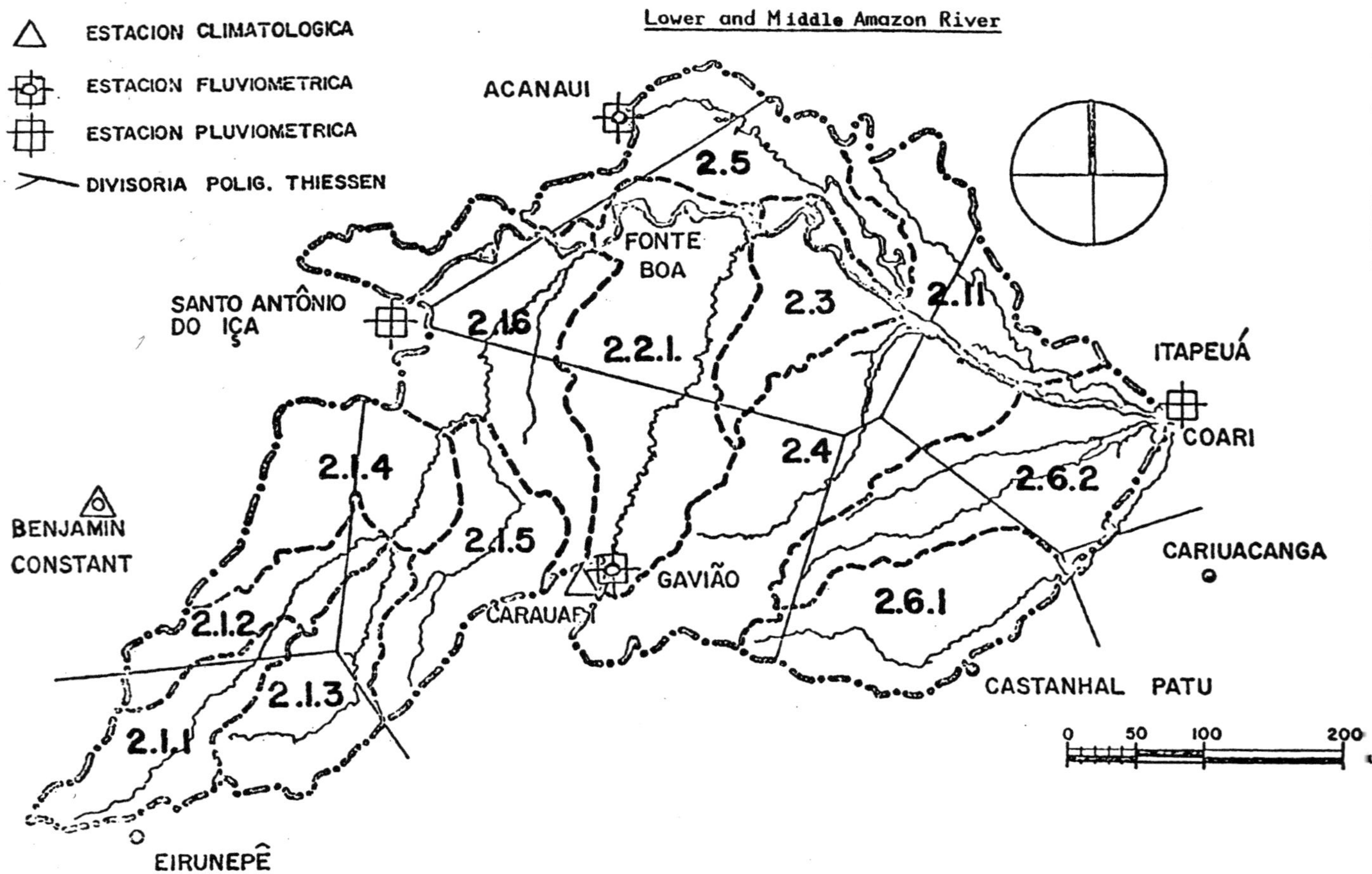

River Solimoes in the basin of the Amazon, modelling sub-bains and polygons of Thiessens.

ANNEX 5 APPENDIX 3

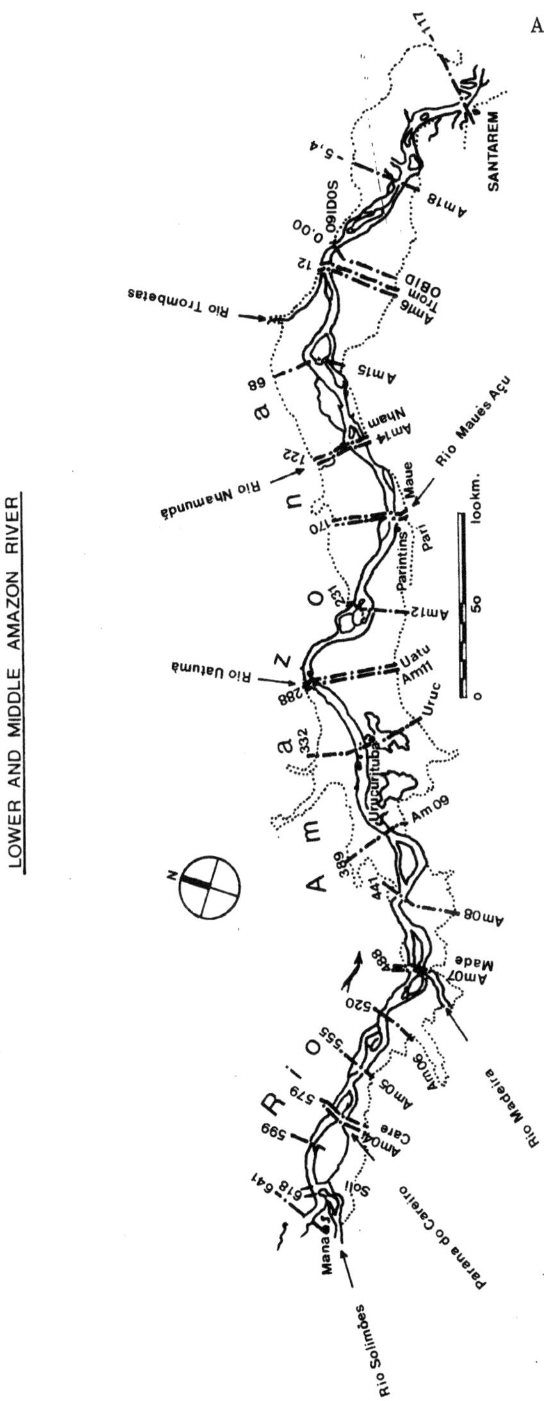

Hydrodynamic forecasts model with profiles of the Lower Amazon.

ANNEX 5 APPENDIX 4

River Tocantins in the Amazon Basin.

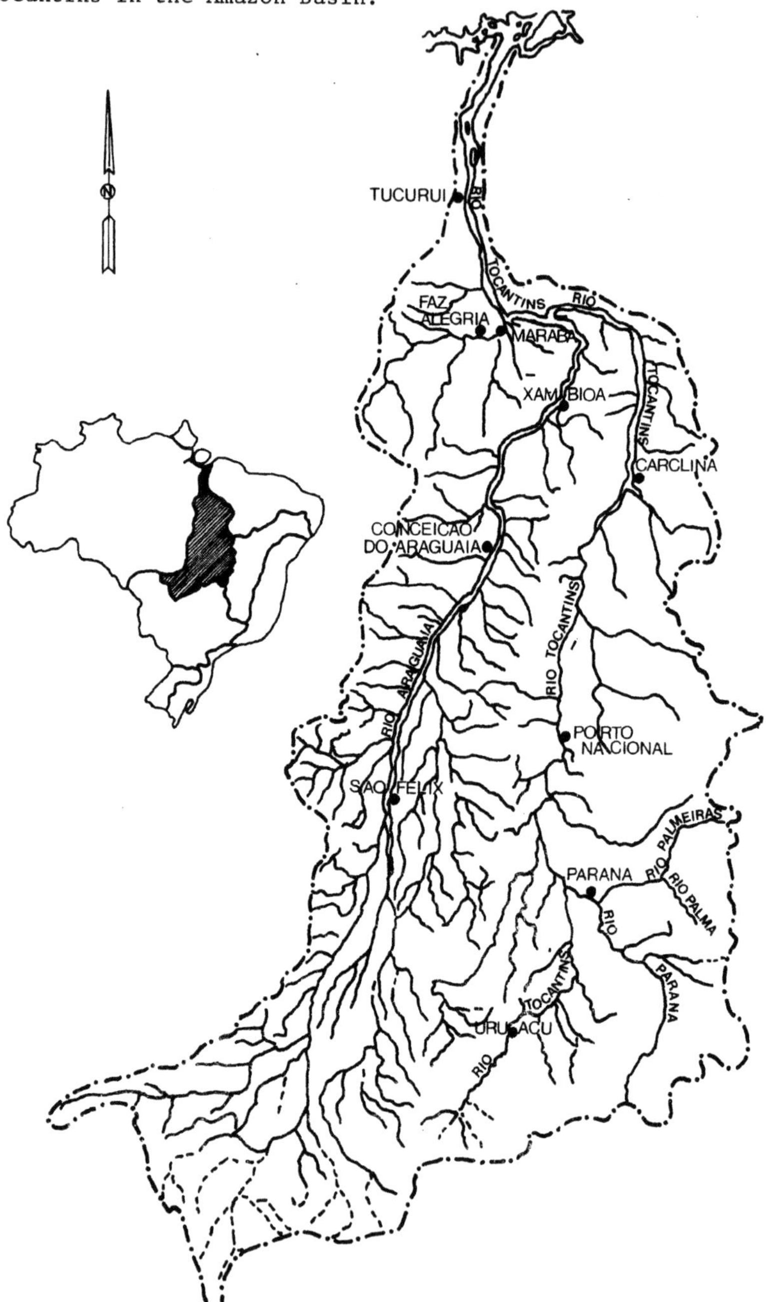

Example of a HFS in the Yangtze River Basin, China

1. COUNTRY: China

2. RIVER BASIN

2.1 Basin name: Yangtze River Basin

2.2 Total area: 1.8 million km^2

2.3 Description of the basin

The Yantze River originates at the border between Tibet and Qinghai province and flows 4300 km eastward to the East China Sea. The climate is subtropical with a basin average rainfall of 1100 mm - ranging from 2200 mm in the lower basin to 250 mm in the west. Seventy percent of the rainfall occurs between May and October. At Datong Station the annual maximum flow of 92,600 cm occurs in August and the annual minimum of 6730 flow at Suhan was 6020 cm in 1923. The basin can be divided into three distinct areas: (1) the area above the Yangtze River Gorges (Three Gorges) is quite mountainous and covers 1 million km^2, (2) the middle reach from the Yangtze River Gorges to Poyang Lake - a distance of about 900 km, and (3) the lower reach below Poyang Lake to its mouth near Shanghai - a distance of 800 km. The lower and middle reach contains nearly 22,000 km^2 of lakes - the largest of which is Poyang Lake with an area of 5000 km^2. The population of the basin is 342 million people of whom 293 million are farmers. The total cultivated area is about 43 million acres. The Yangtze Valley produces nearly 40 percent of the industrial production in China. Five cities in the basin have populations in excess of 1 million people.

3. HFS

A diagram illustrating the system is given in Appendix 1.

3.1 General information

(a) *Purpose of the system*

The ultimate objective is to develop a modern river forecast system for the Yangtze River by:

(i) establishing a pilot automated real-time telemetering system in the Luh River basin;

(ii) establishing a telecommunications link between the Lushui Reservoir near Puche, the Luh River basin data collection center, and Wuhan, the location of the River Forecast Center; and

(iii) improving the flood forecasting center in Wuhan through the introduction of modern rainfall-runoff and flow regulation models.

(b) *History of the system*

In 1931 it is estimated that more than 145,000 people lost their lives during a major Yangtze flood. An even greater flood occurred in 1954 but the loss of life was not as great. Although construction of the Gezhouba Dam will reduce flooding slightly, its total capacity is small. Many major floods have also originated in tributaries of the Yangtze River below that damsite. To achieve maximum efficiency of operation of this major water conservancy project, it will be more important than ever before to provide forecasts of the Yangtze River and tributary flows. Development of an adequate forecasting programme in the Yangtze River will provide early warning to the people living in its flood plain in addition to providing information for navigation and other water-related programmes.

The Gezhouba Dam is currently under construction downstream of the proposed Three Gorges damsite. Its total storage capacity will be 1.58 billion m^3 with an installed hydroelectric capacity of 2,715 mw. This storage represents about one-day flow at the Yichang station. The proposed Three Gorges dam with a capacity of 70 billion m^3 will provide somewhat greater flood control. However, it is important to note that the 1954 Yangtze River flood produced 30 days total flood volume of 138 billion m^3.

Levee systems along the stem have been constructed to more than 1 m above the 1954 discharge values. In addition, several major diversion works have been built to store approximately 50 billion m^3. Some 46,000 reservoirs have been built on tributaries with a total capacity of 114 billion m^3. The largest reservoir is on the Dan Jiang with a capacity of 20.89 billion m^3 and an installed hydroelectric capacity of 900 mw.

(c) *Resources of the forecasting system*

Organization

The system will be managed and operated by the Yangtze Valley Planning Office, in Wuhan, Hubei province, an agency of the Ministry of Water Conservancy.

Computer facilities

A PRIME 750 computer was installed in the Wuhan Forecasting Center.

Operating cost

No figures available.

Capital cost

Land, buildings and capital equipment needed for installation of facilities to be used for telemetry, telecommunications, and processing of hydrological data in process. Cost data not available.

Data-collection system

The pilot project automated the data collection, transmission and forecasting programme for the Lushui River basin above Lushui Reservoir. The Lushui River is located 144 km southwest of Wuhan and is a tributary of the Yangtze. The total drainage area is 3950 km^2 with 3400 km^2 above the reservoir. The basin climate is subtropical with an average annual rainfall of 1600 mm. The annual discharge of the Lushui at Pu Che (immediately below Lushui Reservoir) is 84.1 CMS. Currently, there are 18 rainfall and four streamflow stations in the basin of which 14 existing rainfall and five river and reservoir stage stations will be automated. From Pu Che the data collected by microcomputer andrelayed to Wuhan for processing and issuance of the flow forecasts.

Data transmission system

Currently data from 800-900 rainfall and 100-200 streamflow stations are collected daily. The data are relayed to the center by telephone, radio and/or telegraph. All operations are currently manual. No real-time communication facilities exist at the present time to automatically relay information from Pu Che to Wuhan. Two relay stations will be required to move the data from Pu Che to Wuhan - a distance of 144 km. Data are usually collected about 8 a.m. each day.

Data processing system

The real time data processing has been incorporated in the forecasting procedure.

Receipt and use of meteorological forecasting

Generalized meteorological forecasts products are used during flood situations.

Flood forecasting model

The rainfall/runoff calculations are a modified form of the API (Antecedent Precipitation Index) method and river routing utilizes a modified Muskingum or appropriate procedure. One of the most serious problems involves backwater computations in the many tributaries and lakes in the middle and lower reaches. The forecasts are generally for a three-day period.

4. LIST OF REFERENCES

WMO. Improvement of the Real-Time Telemetering and Flood Forecasting System for the Yangtze River, Request for Assistance under the WMO Voluntary Co-operation Programme HY/3/1/1, 1981.

Flanders, A.F., Consultant's Report on Telemetry and Telecommunications for the Pilot Basin - Part of UNDP/WMO Project Report CPR/80/019, 1981.

ANNEX 6 APPENDIX 1

VHF radio data-collection system for the Luh River Basin.

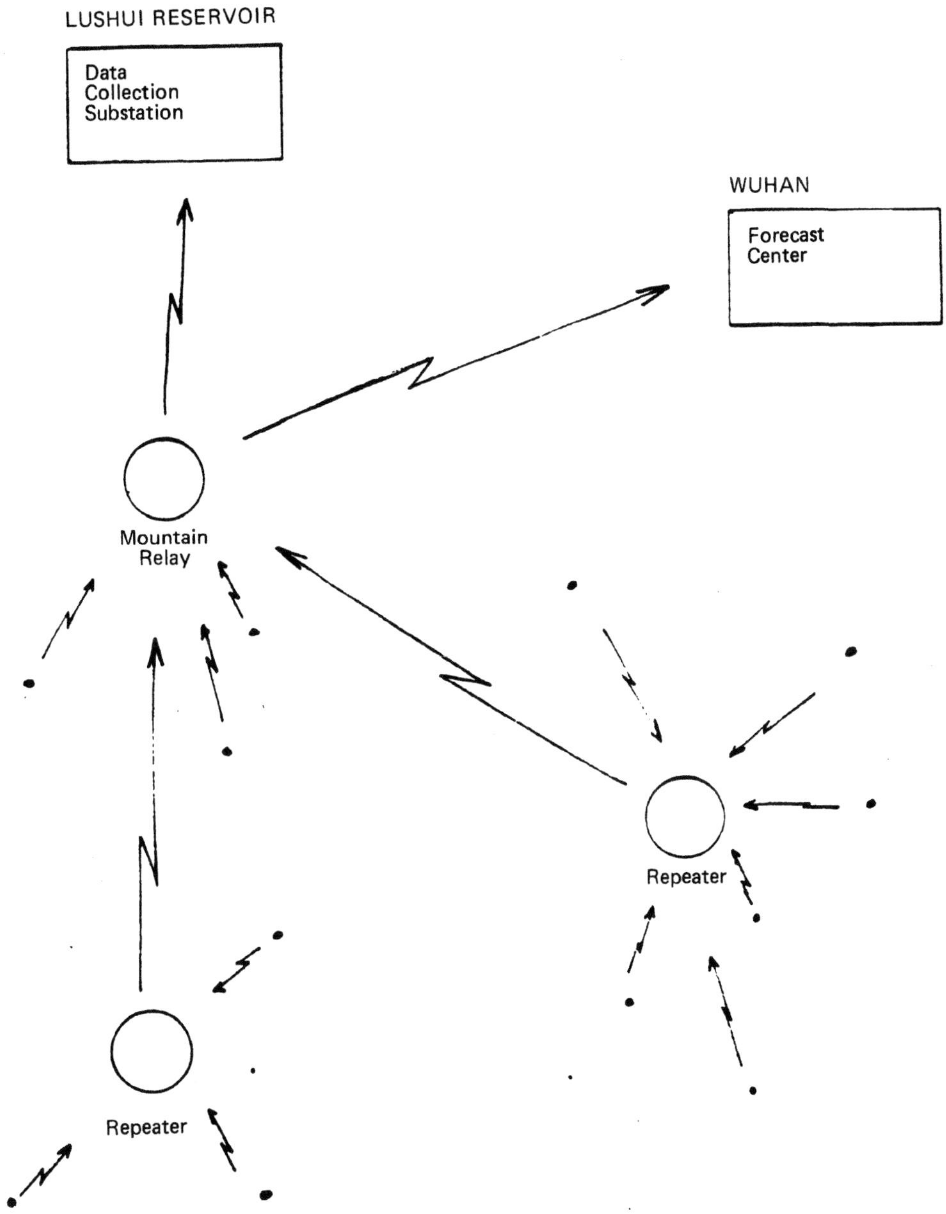

Example of a system installed in and specifications for the data acquisition sub-system for the Niger River Basin

A good example of the advances in the design and operation of a hydrological forecasting system is the basin of the river Niger in Africa. This river, which originates in the mountains of Fouta Djallon in West Africa (Nemec, Manley, 1985), has its basin in nine countries: Guinee, Mali, Niger, Nigeria, Ivory Coast, Burkina Faso (former Upper Volta), Benin, Chad and Cameroon. The length of the river is 4,100 km, the rainfall at the north of the basin area is only 200 mm/year while at the south it reaches 1,500-3,000 mm/year. During its course to the sea it flows through an arid zone of the sub-saharian region, where it forms a so-called, "inner delta", loosing by evaporation in average some 50% of its water. While in the average dry season (September-June) the river is the only source of water for the agriculture of the riparian population, in the flood season it may be devastating, as it was for the capital of Mali in 1968. Thus in 1969 a flood forecasting system was designed for the reaches of the river from its source to the "inner delta" and commissioned for service in 1974. At that time the cost of fully automated data collection stations transmitting by radio was rather high, but this would have not been an unsurmountable problem. The reliability of such an on-line systems was however quite questionable and its complexity, particularly of sensor/encoder/transmitter links, made effective field maintenance difficult. Thus an observer operated SSB radio was preferred to an automatic system.

1. HARDWARE INSTALLED:

1.1 Telecommunication system

At the heart of any forecasting system is the communications network used for the transmission of data and forecasts. The specifications for the bidding process for this telecommunication sub-system are appended to this example.

The reasons for the choice of ARGOS in preference to the Meteor Burst System were as follows:

- ARGOS offered the possibility for each National Forecasting Centre in the eight participating countries to receive its own data, the cost of this with Meteor Burst system would have been prohibitive.

- Power consumption of the data collection platforms (DCPs) was lower with ARGOS.

- The ARGOS transmitters were simple and easier to maintain.

- The experience with the Meteor Burst system has been rather limited for this type of operation.

The installed system consists of 65 Data Collection Platforms (DCPs) and of nine direct receiving stations of which eight will be in the National Forecasting Centres and one at the International Centre at Niamey, Republic of Niger.

The receiving stations are modular for ease of maintenance. Each station has a printer for hard copy output and a data port to allow direct connection to a computer.

Sensors

Forty six of the planned network of 65 DCPs have been installed and are operational.

At each DCP there are following sensors:

- Bubble-type recording water level gauge and direct readout analogue-digital connector which produces the signal for transmission. The compressed air bottle is sufficient for one year normal operation.

- A tipping bucket recording raingauge.

- Sensors for temperature, solar panel voltage remote diagnosis in case of breakdowns.

The DCP has ARGOS radio transmission equipment.

All the equipment except the raingauge and the pipe leading to the pressure take-off in the river, are enclosed in a cabinet which is 2.0 m high, 0.9 m wide and 0.4 deep. The weight is such that it can be moved by a small team of men.

The choice of sites involved a careful study of many possible options. Some of the most important criteria were:

- a well-established river stage/discharge relationship

- interest and utility from a forecasting point of view

- ease of access for construction and for maintenance, particularly important during the rain season

- proximity to a town or village thereby facilitating day to day care for the station.

The installations have not presented any particular difficulty. Two installation teams were used each of which consisted of a professional hydrologist, an electronics engineer and five workers (of which two were drivers). Additional labourers were recruited as required at each site. Local staff who in the future will be responsible for their operation and maintenance also took part in the installation.

There were a few teething problems with the DCPs. Minor problems have been rectified and the platforms have proved to be very reliable with around 97% operational time. The electronics have been completely reliable with all faults being in the electro-mechanical equipment.

Computer facilities

The computer equipment currently used by the Project is a PDP 11/70 mini-computer installed in the buildings of the AGRHYMET Centre, a project of the Sahel Countries, assisting with their agrometeorological and hydrological activities, executed also by WMO. At the moment data are entered in the computer manually, which is not too difficult as only daily data from the 46 existing stations are used.

A micro-computer sytem will be soon connected directly to the central receiving station to allow more efficient use to be made of the data. For example each value of water level received will be used to calculate a true daily mean flow. Graphical presentation of the data will also allow a much closer control on data quality.

Training

A very important aspect of the project for its long-term future has been the training component. This took the form of fellowships for study abroad in hydrology and electronics and group training courses in Niamey.

Software

After competitive tendering the French Company SOGREAH of Grenoble was chosen to develop the forecasting models. They produced a suite of interlinked programs which allowed five modes of operation:

(i) Interrrogation of the data bank. This allows values of the data bank used for model calibration to be studied.

(ii) Calibration of models.

(iii) Model Parameter Bank. This allows the parameters of newly calibrated models to be inserted in the models available.

(iv) Simulated forecasts. Computations using historical data can be carried out with a comparison of recorded and compiled flows.

(v) Forecasting. This is the operational forecasting mode using real-time teletransmitted data.

Four different models are available with the forecasting system (Chene et al, 1985).

(i) CLS (Constrained Linear System). This model uses linear relationships existing between antecedent upstream flows and downstream flows.

(ii) MUSIK. This model is based on the classic Muskingum routing method with calibration by trial and error.

(iii) SAPHARI (Methode Saissonière d'Autocorrélation Multiple pour la Prévision d'Hydrogrammes avec Acutalisation des Résidus-Seasonal Multiple-correlation method for hydrogram forecast with updating of the residuals). The basic assumption of this method is that there is a statistical relationship between the "season" and antecedent-events, and the resulting discharges.

(iv) ARIMA (Autoregressive Integrated Moving Average). This well-known stochastic method has two uses in the forecasting system. The first is to extend the data sequences at the upstream limit of modelled reach and the second is for filling missing flow values.

The computer package uses the "menu" system. It is only necessary to choose a number from a list of available options. The package is also extremely robust, during program testing, attempts were made to crash the system by giving out-of-range replies or replying with a station name rather than number. Never once did the system fail. A separate model which simulates the inner delta of the Niger River in some detail is being prepared by ORSTOM under a separate contract with the NBA.

2. FIRST RESULTS

The first test runs of the models using teletransmitted data were carried out in February 1985. Regular use of the model has been carried out since April 1985, though at that stage data from a number of key stations were not yet available in real-time so the forecast coverage has been incomplete.

The system has proved itself to be easy to use, about 30 minutes being required to run the data preparation program (once the data has been entered), the modelling system and output the results. It is however too early to comment in detail on the accuracy of the results. The model can produce forecasts with a lead time up to three months, however, during the early part of the annual flood the predictions are heavily influenced by rainfall, which at present is not taken into account in the models.

3. PLANNED SECOND PHASE OF HYDRONIGER

The first phase of the project comes to an end in December 1985. A second phase, starting in January 1986, is planned. The main activities of this phase will be:

- To instal additional DCPs. This will particularly aid some of the countries not on the main river in the upper basin area for which forecasts are not at present available.

- To instal additional raingauges and develop a rainfall/runoff model. This will permit upstream forecasting and will improve forecast accuracy at locations where runoff from ungauged intermediate catchments provides a significant flow contribution.

- To improve computing facilities in terms of hardware and software particularly with regard to graphics output and quality control of data.

- To work closely with participating countries to provide effective preventive maintenance and efficient repair of breakdowns in the system.

- To improve methods of communicating forecasts to the users.

Other possibilities under active consideration are the addition of extra sensors (i.e. turbidity of water) to existing DCPs and the use of remote sensing by satellite to improve real-time area rainfall estimates.

APPENDIX

Specifications for the data acquisition system for the Niger River Basin

1. PREFACE

The real-time hydrological data acquisition system will measure the water stage at selected points located on the River Niger and principal tributaries and transmit these stage readings to the International Forecasting Centre (IFC) located in Niamey, Republic of Niger. At the IFC, stage readings will be recorded on magnetic media and be printed out on hard copy for visual examination. The complete operating system will consist of remote Data Collection Platforms (DCPs) installed at 95 remote sites and a central receiving station including a communication controller and processor, a magnetic media recording device and a printer that will be installed at the IFC. The data acquisition system will be operated at least once a day in normal operation. The interruption of outstations between transmission is desirable but not indispensable. The system as a whole should reliably operate for more than 95% of time.

Provision will be made to disseminate the data obtained at the IFC and resulting forecasts to the National Forecasting Centres (NFCs), if the proposed data collection system would not provide for such dissemination. In this latter case, data and forecasts will be disseminated through existing PTT facilities, reinforced ultimately with an appropriate communications system from the IFC to the NFCs.

2. TECHNICAL REQUIREMENTS

2.1 Remote station locations

The DCPs will be located at various points within the boundaries of the eight West African nations which contain the watershed area of the River Niger. These nations are: BENIN, BURKINA FASO, CAMEROON, GUINEA, IVORY COAST, MALI, NIGER, and NIGERIA. The platforms will be installed along the River Niger and some of the tributaries of the Niger. The exact remote site locations are given by latitude and longitude co-ordinates as well as the measured distance upstream from the mouth of the river on which it is installed (see tables), along with sensor performance requirements given below.

2.2 Description of measuring conditions

2.2.3 *The range of water levels*

Water level observation data are available, in series of varying lengths, for almost all the foreseen data collection sites. With regard to these, the sites can roughly be classified according to their range of water levels attaining 5, 10 or 15 meters. The number of sites within each range is estimated as follows:

(a) Range attaining, but not exceeding 5 meters:

21 sites - 22% of all the sites;

(b) Range attaining, but not exceeding 10 meters:

46 sites - 48% of all the sites;

(c) Range attaining, but not exceeding 15 meters:

28 sites - 30% of all the sites.

The above figures should only be considered as of informative value giving an indication of measuring conditions. Allowing also for unforseeable circumstances, the maximum ever range should be put at 20 meters.

2.2.4 *Conditions concerning the safety of instruments*

The security of the instruments installed at the measuring sites will require the employment at each site of a "station master" whose task will be to guard the instruments from all possible infliction of damages, either by humans or by animals. The "station master" will also remove dust from the solar panel and perform similar duties. Natural conditions may, however, present problems for the proper functioning of the measuring point - against which the "station masters" can hardly do much. Whatever the type of connexion between the instrument and the sensor (closed or open channel, tube carrying liquid or air, electrical cable) the security of the sensor may be endangered by river scouring. Furthermore, in the case of water conduits, clogging is an apparent danger. Thus, the stable and safe positioning of the measuring point, i.e. the orifice of any type of conduits, or the sensoring device of the instrument, will almost certainly pose problems at most of the sites where the effects of river scouring and depositing have to be taken into account.

2.3 Requirements concerning DCP instruments

By considering the above general description of the conditions among which the DCPs have to function, the requirements that their water level sensors and recorders have to satisfy can be formulated as follows:

2.2.1 *Climatic conditions*

Climatic conditions are very variable throughout the River Niger Basin. It is a requirement that each instrument should be able to function under the climatic conditions of any part of the basin and therefore the extreme values should be taken into account for each station. The following conditions and values are considered relevant:

(a) *Extreme temperatures* - Maximum air temperatures may reach 50°C and minimum temperatures may drop to 0°C. Diurnal variations result in temperature differences which may exceed 25°C within an interval of a few hours only.

(b) *Extreme air humidities* - Relative air humidity may vary between extreme values within the year, even at the same location. In general, one should count with a minimum value of about 10% and with a maximum value of 100%, with the occurrence of condensation.

(c) *Extreme weather conditions* - The occurrence of thunderstorms, of heavy rainfalls and of gale-force winds should be reckoned with. Furthermore, sand-storms are quite common over vast parts of the basin, when very fine particles of solid material are carried by the air that can be excluded only from hermetically sealed boxes.

2.2.2 *Topographical and hydraulic conditions*

The data collection sites will be along rivers which may vary in their size, but which are almost invariably in unregulated conditions. This implies that:

(a) Access to the site might be difficult in part of, or throughout the year.

(b) Rugged surface and/or dense vegetation should be reckoned with at the river bank.

(c) At some places, the river's low water bank and high water bank are separated by distances which may be in the range of kilometers. As a result measuring instruments may have to be placed at a considerable distance from the main course of the river or may have to be perched on a platform situated above the highest flood levels.

(d) The possibility of an unstable, moving river bed exists. The scouring and sediment depositing conditions may be such as to endanger the stability and/or the entrance of the water to the measuring point.

2.3.1 The operational vertical measuring range should be 20 meters, and the measuring accuracy should be one centimeter. The former requirement is for the sake of having interchangeable instrumentation. However, if major economies and/or technical simplifications warrant it, separate groups of instruments may be considered with different measuring ranges. Even so, not more than three groups are recommended with 5, 10 and 20 meters of range.

2.3.2 The instrument should be fully operational within an open air temperature range from 0°C to 50°C. Temperatures developing within the housing can reach 60°C. The encoder and electronics in the housing must be fully operational in the temperature range 0°C to 50°C, and it is desirable that they be able to operate in a temperature range environment higher than this.

2.3.3 The instrument should be fully operational even under extreme humidity conditions and amid the roughest weather conditions; therefore, each system should be in a hermetically sealed container including a dessicant such as silica-gel. It is also preferable that this container be practically shock-proof and be as robust as possible. The antenna should, if possible, be protected by a dome or shelter.

2.3.4 The instrument should operate for long periods without attendance and further, this attendance should require as few technical skills as possible.

2.3.5 The connexion between the instrument and its measuring point should be robust enough to withstand the effects involved with moving river beds. Also, the greater the maximum permissable length of this connexion, the more advantageous it should be considered - versatile applicability to situations where the location of the instrument and that of its measuring point are necessarily distant from each other.

2.3.6 The instrument's measuring point - the orifice of a conduit or a sensoring device - should be the least susceptible to damages caused by river scouring and sediment depositing.

In summary, sensors utilizing any kind of measuring techniques which satisfy the above requirements are acceptable for the DCPs.

2.4 DCP interface, power and transmission equipment

2.4.1 Perfect compatibility between parts of the DCP should be assured. Joining of components as well as to the connexion to the measuring point should be realized in the most easy-to-handle way, all inter-connexions utilizing standard configured connectors, in such a way that maintenance can be achieved through replacement of modules.

2.4.2 If a satellite is to be used in the data transmission system, the transmitted message format, power output, antenna, modulation, coding and carrier frequency shall be in accordance with the appropriate specifications issued by the satellite operating agency. The equipment shall be certified by the satellite-operating agency. If a system other than a satellite-based system is to be used, the platform transmitter terminal shall use a combination of RF power output and antenna configuration such that data can be reliably transmitted up to 2,000 kilometers to the IFC in any direction from the platform site. Adequate means shall be provided to ensure the accuracy and reliability of reported data. The possibility of interrogating individual DCPs would provide such a mean.

2.4.3 The station I.D. number and repetition rate of transmission, if not an interrogation system, shall be established by pre-wired plug-in modules or a plug-in jumper arrangement, or by a programmable system on the station. A portable or stationary test set should check:

- the power consumption;
- the operating cycle;
- the transmitted message;
- the programme parameters if the station is equipped with a microprocessor.

2.4.4 The initial DCP installatin will be for the purpose of transmitting water level and battery power information. However, the number of transmitting parameters may be increased at a later phase (e.g. precipitation data). Therefore equipment design should allow for transmission of not less than three parameters with sufficient accuracy (16 bits) without modification.

2.4.5 The power supply for each DCP shall be a re-chargeable battery. The battery shall be of sealed cell, of maintenance-free design and thus eliminate the need for checking electrolyte level or the addition of distilled water or other liquids to maintain the desired electrolyte level. The capacity of the battery shall be sufficient to operate the DCP equipment for five days without loss of performance of any function with zero input from the charging source. The battery shall be charged from a solar panel assembly consisting of a solar-cell array hermetically sealed to prevent entry of dust, moisture or pollutants, a voltage regulator to maintain the correct battery charging voltage, low voltage cut-out protection and a blocking diode to prevent reverse current from discharging the battery during no-sunlight periods. The solar array shall be sized to have a capacity such that when connected to a battery having only 25% of its Ampere/hour capacity remaining, and operating a normally functioning DCP, the battery will be fully recharged within 16 hours when the solar array is illuminated by the unobscured sum moving through a 120°arc (zenith ± 60°). An allowance is to be made for the possibility of some dust settling on the solar panels.

2.4.6 The DCP equipment, excluding the solar panel and water level sensor, may be housed as described below. The electronics equipment and battery shall be contained in a strong, lockable, pre-fabricated metal structure or container mounted on a firm, stable base. The battery shall be installed in a compartment sealed and isolated from the compartment containing the electronics equipment. The antenna for the station may be mounted on top of the structure or container or on a mast located close to the station. If a directional antenna is used, it will be equipped with a base which can be adjusted in azimuth and slope. All external connexions to the solar panel, sensor and antenna, if applicable, shall be by means of waterproof cable entrance bushings containing a compressed neoprene insert or an equivalent fitting. Consideration should be given to mounting equipment in an easily transportable container capable of withstanding shock in movement, and which on arrival at location may be easily mounted or installed by semi-skilled personnel. Due to variable installation requirements, the DCP housing should be suitable for mounting on a 3 to 5 meter tower type structure without degradation of environmental sealing. Specifically, all conditions described in paragraphs 2 and 3 above must be satisfactorily met.

2.4.7 Specific requirements for the station infrastructure, that is a standard plan for arrangement, mounting and installation should be provided by the contractor in his proposal.

4.8 Additonal aspects to be considered concerning the infrastructure and equipment will be safety, imperviousness, locking device and corrosion proofing. Any ventilation openings shall be fitted with louvers to prevent entrance of rain and screened to prevent entrance of insects, small animals and reptiles. Given the high temperatures, it seems that double-roofed shelters are indispensable.

2.5 Central station equipment - Niamey (Niger)

The central station equipment at the IFC will consist of receiver(s) to receive the remote data, a mini-computer to serve as communications controller and data processor, a hard copy printer, a magnetic media recorder, air conditioning and an uninterruptable power supply with 15 minute power failure capability. A suitable diesel generator must be provided for in the proposal which will adequately power all station equipments. An option in the proposal will provide for a second back-up generator. Available commercial power supply is at 220 Volt 50 Hz, but variations of ± 30 Volt and ± 2 Hz around the nominal values should be accounted with.

2.5.1 Receiving equipment shall provide adequate sensitivity to reliably receive and demodulate remote station transmissions in any direction from the receiver site. Signals shall be amplified and the extracted data shall be properly fed to the communications mini-computer. If a satellite system is proposed which utilizes more than one frequency, provision will be made to receive all employed frequencies. Appropriate facilities should be provided for testing the proper functioning of the electronics and to determine any possible failure of the station.

2.5.2 The communications processor will automatically check the digital data from the receiver for correct message format, validity and I.D. It will append the Julian data and 24-hour time (minutes explicit) to the data if valid. Valid data will be recorded. Invalid or illogical data will be so annotated. The communications processor will have two outputs for this data. One output will go to a magnetic media recorder, possibly a cassette or diskette recorder. The second output will be to a hard copy printer (at least 60 cps) for visual examination to show the functioning of the system and river stage readings. It will be an advantage if the communications processor can be accessed from another computer through standard communication interfaces (RS-232 and 20mA current loop).

2.6 System reliability

The system will be utilized to acquire, transmit, evaluate and disseminate vital data affecting the safety of life and property. No specific tolerances concerning specific equipment or methods of processing are given herein, but proposals shall take into account the requirement for accurate and timely data transmission and provide information on overall system performance and reliability.

2.7 Data dissemination system

Data received from the remote stations, as well as forecasts based on the same, will be transmitted from the IFC to the NFCs situated in the member countries of the NBA, probably in the capitals, if this is not integrated into the data collection system. Information concerning operational status of remote stations may also be transmitted. It is foreseen that this transmission will be by means of existing PTT facilities. Also, it is possible that this system should be reinforced. The contractor's proposal should include as an option the provision of an appropriate transmitter equipment at the IFC. This equipment should include a transmitter system with omni-directional antenna, frequency shift keyer, remote control operation, adequate spares and test equipment. Provision will be made for transmission on several frequencies. To ensure reliable transmission, a duplicate system will be provided.

2.8. Supplies and spare parts

In addition to all equipment required for the installation and activation of this system, the following supplies and spare parts shall be provided:

2.8.1 Instruction Manuals and parts lists will be provided for each station in French and English.

2.8.2. Hard copy printer paper and printer tape or ribbons and other supplies for the central station and the DCPs - sufficient for two years of operation.

2.8.3 Magnetic Media Recording Unit. Sufficient recording material for two years should be provided.

2.8.4 Twenty (20) spare complete electronic systems for the remote data collection platforms complete with programmable plugs or modules to set station I.D. numbers and transmission intervals, if such units are used in the operating stations.

2.8.5 Ten (10) spare solar panel assemblies.

2.8.6 Twenty (20) spare remote station battery units.

2.8.7 Ten (10) battery chargers operating from 220-volt, 50-cycle mains to maintain the spare batteries at charge. Chargers will be capable of rapid charge and trickle charge.

2.8.8 Ten (10) test sets for checking the DCPs.

3. MAINTENANCE

3.1. The contractor for this system shall also provide maintenance for all remote stations and the IFC equipment for a period of one year with options for renewal for five (5) additional years. The contract price for the additional years, if this option is exercised, shall be established by a previously mutually agreed upon formula, such as the initial year contract price increased by the percentage increase of the cost of living in the region.

3.2. The contractor shall provide equipment, tools and material to allow establishment of a central workshop at the IFC in Niamey. Provision will be made for all appropriate special test equipment, electronic technical tools and consumable maintenance material as well as a stock of those discrete components which are most likely to fail. Storage cabinets, work benches and similar equipment will be included. The workshop should provide for the full maintenance of the instruments supplied. This item will be evaluated separately.

3. Equipment repair should be performed in the IFC central workshop once available.

4. TRAINING

4.1 The contractor shall provide technical training instruction at the IFC for a period of not less than one month to train at least ten national electronic technical personnel on circuit theory, operation and maintenance of all the equipment provided.

4.2 Prior to the acceptance of the equipment, the contractor will provide advanced training on software and electronic circuitry for two technicians designated by WMO, for a period of approximately two weeks.

5. DELIVERY

5.1 The proposal should provide for delivery and installation of all equipment within no more than 18 months after award of contract.

5.2 An option for the proposal would be to allow delivery to be made in three phases as given below:

(a) Thirty remote stations and the central station (IFC) to be installed 12 months after award of contract.

(b) Thirty additional stations and the IFC workshop to be installed 24 months after award of contract.

(c) Remaining 35 remote stations and the data dissemination system to be installed 36 months after award of contract. This phase of the proposal should be quoted as an option.

6. GUARANTEE AND ACCEPTANCE

6.1 All equipment will be tested at the factory or an appropriate facility through a burn-in test that will be carried out while the electronics are operating and will last at least 72 hours during which the temperature will range from 0° to 60°C as follows:

- 2 hours levelled off at 0°C;
- increasing at a rate of 10°C per hour up to 60°C;
- 2 hours levelled off at 60°C;
- decreasing at a rate of 10°C per hour down to 0°C;
- 2 hours levelled off at 0°C, etc.

The relative humidity will range from 0 to 90%. The equipment should perform according to specification during and after this burn-in test. WMO and the project representatives shall be permitted to attend these tests.

6.2 WMO reserves the right to make additional tests under tropical conditions and to have the equipment performance checked by its own experts.

6.3 All equipment will be installed and checked by the contractor before final turn-over and acceptance by the project's official representatives.

6.4 The guarantee period for all equipment will be not less than one year after acceptance by the project's representatives. The contractor will indicate the warranty period and procedures.

Jhelum district flood order 1980
Anti-flood measures/relief work in Jhelum district

General

1. This order shall be called Jhelum District Flood Order 1980. It cancels all previous orders and will come into operation on receiving high flood warning for river Jhelum.

2. Jhelum District was extensively affected by floods by river Jhelum prior to the construction of Mangla Dam. After the construction of Mangla Dam, the release of water in the river is controlled and frequency of flood has been minimized to some extent. The villages affected are situated on the Western Bank of river Jhelum, right from Mangla in the North down to Ahmadabad in South of the District. The damage due to flood is further aggravated by torrential rainy Nullas which appear to be fairly dry and innocent during dry season. This District was affected by river floods and hill torrents in 1973 and again in 1976.

District Flood Relief Committee

3. Following District Flood Relief Committee is hereby constituted:

a. Deputy Commission (Chairman)
b. Brigade Commander 42 Arty Independent
c. Brigade Jhelum or his representative
d. Superintendent of Police Jhelum
e. A.D.C. (G)
f. Assistant Commissioner Jhelum
g. Assistant Commissioner Chakwal
h. Assistant Commissioner Pind Dedan Khan
i. Executive Engineer Irrigation Jhelum
j. Executive Engineer Highways Division Jhelum
k. Executive Engineer Provincial Buildings Division Jhelum
l. Executive Officer Cantonment Board Jhelum
m. EAC(R)/District Flood Relief Officer (Secretary)
n. Medical Superintendent District Headquarters Hospital Jhelum
o. District Health Officer Jhelum
p. District Food Controller Jhelum
q. Chairman Municipal Committee Jhelum
r. Civil Defence Officer Jhelum
s. Assistant Director Social Welfare Jhelum
t. Assistant Director Live Stock and Dairy Development Jhelum

Classification of flood

4. Following is the classification of flood in river Jhelum:

a.	Low flood	0.75 to 1.10 lacs cusecs
b.	Medium flood	1.10 to 1.50 " "
c.	High flood	1.50 to 2.25 " "
d.	Very high flood	2.25 to 3.00 " "
e.	Exceptionally high flood	3.00 lacs cusecs and above

Flood warnings and means of information

5. Advance information regarding flood in river Jhelum will be communicated by Resident Engineer, Reservoir, Mangla and Executive Engineer, Rasool Headworks for discharges at Mangla and Rasool respectively through wireless/teleprinter/telephones. The discharges of hill torrents crossing Upper Jhelum Canal into river Jhelum at Jatli, Rahmanpur and Jaggu will be intimated by Head Signalman, Jhelum (who is under Executive Engineer, Irrigation Jhelum) through telephone No. 2928 or through a special messenger. Wireless sets will be immediately located at Mangla, Jhelum and Rasool.

6. Flood warning is also received from the Provincial Flood Warning Centre Lahore, through teleprinter.

7. The flood warning will immediately be communicated to the Deputy Commissioner, Superintendent of Police, Assistant Commissioners and all other Sector Officers. The Sector Officers will convey warning to the population likely to be affected by flood. No warning will be issued for low or medium flood unless there are immedaite prospects for high flood. However, daily discharge of water at both the vulnerable points will be plotted on board in Flood Control Centre.

8. High flood message should immediately on receipt be repeated either by telegram or by wireless to to the Assistant Commissioner, Pind Dedan Khan, SHO, Jalapur SHO Lilla, Station Master Haranpur, Station Master Chalisa and SHO Dina and to all Sector Officers. All subsequent telegrams showing rise or fall of water will be repeated to them. The telegram will read in the following terms "Jhelum river in high flood"(.) Discharge ____________ cusecs (.) warn all concerned(.) Be prepared to tackle the situation".

9. On receipt of high flood warning exceeding three lacs cusecs, all schools for girls and boys in the affected areas shall be closed down under the orders of the ADC(G) or Sector Officers.

Flood warning for Jhelum town

10. In Jhelum town high flood warning will be issued by the Civil Defence Officer through sirens. No siren will be necessary for low or medium flood. A discharge of more than 2.25 lac cusecs of water will be regarded as a very high flood and flood warnings will be given through the loudspeakers. Flood warnings will also be issued through the microphones of Jamis Masjid Nawan Mohallah, Gumbadwali Masjid, Masjid Hafiz Noor Mahd,Masjid Noman and Masjid Khansaman Nia Bazar. One official will be posted in each of these mosques who will get information regarding the flood from the District Control Officer and will issue the same through the microphone of the mosques. Chairman Municipal Committee Jhelum will detail five officials of the Committee immediately on receipt of the high flood warning for duty in the mosques mentioned above.

11. On receipt of warning of the discharge of water above 2.25 lacs cusecs at Mangla, the army authorities will be requested to be on one hour's notice to render help to the affected areas. The army authorities shall be requested to despatch help immediatly by Deputy Commissioner whenever the situation so warrants.

District Flood Control Centre

12. There will be District Control Centre at the Headquarters in the District Courts and it will function round the clock during the flood season. The FAC(R)/District: Flood Relief Officer will be in charge of the Control Centre. He will be assisted by the Chief Officer Committee Jhelum, Chief Officer Zila Council and Superintendent Deputy Commissioner's Office. The Head Clerk, a typist and two peons will be on duty at the Control Centre. On receipt of high flood warning the District Nazir and HVC, will also be on duty at the Control Centre. A map prepared by the revenue staff showing the areas affected by flood during the post shall be available in the Control Centre. It will also indicate sectors and sector officers.

Jhelum Tehsil - Jhelum Town

13. Jhelum is a town which has perpetually been subject to flood ravages, Mohallas like Piragheb, Grain Market, Nia Bazar, River Road, Tensil Vicinity etc. are often affected, requiring evacuations.

14. There is Nullah known as Kala Kassi near the old hospital which is a source of constant danger both to the hospital and the residents of the locality. It draws water from the river and also collects rainy water from a vast area below Rothian and spreads in the low lying locality of Jhelum city on both sides of the Civil Lines Road. Similarly below the river bridge, there are big drains

which are choked and flood the roads between the Railway Colony, Railway Road, Railway Bridge and Paradise Cinema. The Railway Department and the Municipal Committee will make adequate arrangements for the drainage of this water by means of pumps. The Sector Offices of Jhelum Town and Chairman Municipal Committee are required to be vigilant in regard to these areas. Sufficient pumps are to requisitioned immediately so that they can be moved to affected areas within half an hour.

Rural area

15. Apart from Jhelum Town. Villages of the tehsil situated on river belt are also affected by flood in river Jhelum and by hill torrents. These are at Appendix "A".

Sectors - Sub-sectors and officers responsible

6. Jhelum Town. The Assistant Commissioner Saddar will be overall in charge of the Jhelum Town. He will be assisted by the ASP/DSP (Headquarters) and Chairman Mnunicipal Committee Jhelum. The town will be sub-divided into following two sub-sectors.

Sl/No.	Name of Sector indicating its area	Name of Sector Officer and location of his office
a.	Northern sector	
	This sector will cover area of Jhelum Town located on the Northern side of Machine Mohalla Road, Shahdar Chowk, Tehsil Road	City Magistrate assisted by Municipal Committee Jhelum Office in Municipal Committee Telephone No. 4082
b.	Southern sector	
	This sector will cover area of Jhelum Town located on the southern side of Machine Mohallah Road, Shandar Chowk, Tehsil Road	Tehsildar, Jhelum assisted by Civil Defence Officer, Jhelum Office at Tehsil HQ Jhelum Telephone No. 4042
c.	Rural area	
1.	Sheikhupura down to Pahhwal Dangerours areas are Sheikhupura and Vichla Bela	General Asstt; to D.C. assisted by SHO Dina and Circle Kanungo
2.	Tawakalpur Pakhwal to Kahan Nullah. Dangerous areas are Pakhwal and Naugran	Mr. Karam Dad Special Traffic Magistrate assisted by Tehaildar Jhelum

3. Kahan Nullah to Jalalpur	Asstt; Director Local Govt assisted by C.O. Zilla Council, Circle Kanungo and a Separate Police contingent of one platoon detailed by the Supdt. of Police

Note: The Office Kanunga will be responsible for the safety of all the Revenue Record held at the Tehsil Office. Moharrir Registration Tehsil Qasids shall report to Tehsil Office immediately after the siren and will assist the Office Kanungo in safety of Office record.

Pind Dadan Khan Tehsil

17. This sub-division is particularly vulnerable to flood especially at PD Khan proper and village Kanianwal. There is a ban maintained by the Town Committee for the protection of the town. The Asstt. Commissioner and Chairman Town Committee with the voluntary help of the labour of Municipal Committee, Soda Company, Salt Mines and Cement Factory will make adequate security measures for its maintenance immediately.

18. The Executive Engineer Highways shall maintain roads leading to Khawra and Jhelum too and shall make arrangements for putting up demarcation marks on all roads and streets submerged under water for the guidance of the traffic and put up closure notice boards as dangerous because of an old tank there. The flood water on the road invariably joins up with the tank water and thus forms a big lake.

19. The SDO Highways Chakwal and PD Khan shall take special measures to maintain the hill road between Chan Saidan Shah and Khawra in perfect order.

Sector - Sub-sector of Pind Dadan Khan

S/No.	Name of Sector indicating its area	Name of Sector Officer and location of his office
a.	PD Khan Town	Assistant Commissioner PD Khan
b.	Jalalpur to Azampur	Project Manager PD Khan assisted by SHO Jalalpur & Circle Kanungo, Office Jalalpur
c.	Azampur to Pithar Kalan Sial to Ahmadabad	Tehsildar, PD Khan assisted by SHO Lilla Circle Kanungo

Chakwal sub-division

20. Chakwal Tehsil is also affected by some Hill Torrents. Asstt: Commissioner Chakwal will be Sector Officer for Chakwal Sub-Division and will be responsible for undertaking precautionary measures/relief operations within his jurisdiction as indicated in the overal framework of the scheme for other sub-divisions.

21. All Sector Officers must reach their stations immediately on receipt of a high flood warning. The Project Managers Markiz, Lamberdars, Public representatives, kanungos, Patwaris and the supervisors and workers of the area will be at the disposal of the officer on duty. These officers/officials will be assisted by the representatives of Food, Agricultural, Veterinary, Medical, Highways, Buildings and Irrigation Department. The sector officers shall form Flood Relief Committee in their sectors. The Sector Officer on reaching the sector shall ensure that necessary information is conveyed to the villagers. He should prepare a list of volunteers and have with him a map showing the area of his sector. Maximum participation of the local counsellors and other soil welfare organizations should be ensured.

22. The Sector Officer will collect boats of the area in most suitable places and will render all possible assistance to the flood affected villagers. He will shift the population of the affected area to the nearest Relief Camp if required. He shall also arrange supplies of atta and other staples with the help of the Food Department. Medical cover will be provided by the District Health Officer and the Assistant Director Livestock and Dairy Development. As a precautionary measure immediate inoculation/vaccination of all population in the District be carried out by the District Health Officer.

23. The Sector Officer must remain in constant touch with the District Control Office at the Headquarters and send daily diaries to the Headquarters of the Casualties of persons and cattle and damage to the property. In case he is unable to cope with the situation he will immediately inform the Deputy Commissioner about it stating the nature of help required. The Sector Officer will send daily flood situation report.

24. As soon as the flood recedes the sector officers should submit a report to the Deputy Commissioner giving the exact course of the flood, its height and the damage done to the villages indicating loss of human lives, property and damage to cattle crops etc. He will also attach a map with the report showing the course of the flood.

Arrangement of boats, trucks, etc. for evacuation purpose

25. Sector Officers/Sub-Sector Officers will ensure that required number of boats/trucks are requisitions well in advance and are available with them. They would be responsible for providing trucks/boats to the affected areas without loss of time. Mulahs will be employed on daily wages.

26. SHO City Jhelum will make available maximum number of pneumatic tubes through voluntary contribution by Transport Companies of City and will hand them over Sector Officers immediately after the receipt of high flood warning at Municipal Office for evacuation purposes.

27. Chief Officer Zila Council shall be responsible for making arrangements for boats when and where necessary outside the town. The services of local boatman should be utilized on daily wages.

28. The Tehsildar PD Khan should arrange for evacuation etc. and keep available at least ten boats with crews on duty will ensure that at least two boats are kept ready within their respective areas. Any number of spare boats from PD Khan or the adjoining ferries should be requisitioned whenever required.

29. The evacuation of the affected population will be made by the Sector Officers from respective areas under their personal supervision.

30. The Sector Officers/Sub-sector Officers shall be meeting in their respective areas immediately on receipt of this order and shall form a committee consisting of suitable residents of the Ilaqa and officials placed at their disposal, who shall report to them immediately on receipt of the High Flood Warning and render all possible help to them.

Police help

31. The SHOs of the flood affected areas shall assist the Sector Officers in the discharge of their flood duty.

Medical aid

32. The Medical Superintendent, District headquarters Hospital Jhelum with the assistance of Municipal authorities will provide medical cover to the urban area and the Relief Camps in flood affected areas within Municipal Limits. The Distt:Health Officer will arrange medical relief to the Relief Camps in flood affected rural areas. He will make sure that adequate stock of vaccine, drugs, antibiotics and anti-snake bite serum are available in each sector. All curative and preventive measures are to be taken by the

Health Personnel and Municipal Health Authorities along with Zila Council sanitation staff in city and rural areas.

33. The Assistant Director Livestock and Dairy Development will ensure that all the veterinary Officers/Stock Assistants in the flood affected areas have adequate stocks of medicines before the commencement of flood season. The V.Cs shall provide medical aid to the animals affected by flood as soon as the flood water recedes.

He should also arrange to provide the medical aid to the Relief Camps as and when required.

Food supplies

34. District Food Controller Jhelum will ensure that adequate stores of food grains are available with him. He will be responsible for making supplies to the flood affected area and relief camps established by the Distt: Administration. In case of shortage of food grains, he should immediately make arrangements for replenishment of his stock through his department.

Fodder

35. Deputy Director Agriculture shall be responsible for supplying dry fodder for cattle in the affected areas. He will ensure that stock lying in the low-lying areas are removed to safer places at short notice. He will utilize the services of his own staff and local volunteers for this purpose.

Control store for relief articles

36. A Central Store shall be maintained for collection and storage of relief articles donated by various individuals/organizations. Assistant Director Social Welfare will form a Committee for collection of donations/relief articles. He will be in charge of store. He will be responsible for proper distribution in the light of instructions issued by the Deputy Commissioner. Cash donations will be doposited in the District Relief Fund to be operated by the Deputy Commissioner.

Relief camps

37. Relief camps have been earmarked and will start functioning as and when necessary. The list is given in Appendix B.

Post-flood measures

38. As soon as the flood recedes, relief work in the form of

distribution of amenities, tacavi loans and gratuitous relief will be taken in hand. Assistant Commissioners shall visit the affected areas and shall be personally responsible for assessing damages correctly. The Chairman Municipal Committee Jhelum shall assess the damages caused to houses etc. in City Jhelum.

Tacavi loans

39. The Assistant Commissioners will immediately send their approximate requirements of tacavi loans and ex-gratis grants as laid down by the Board of Revenue.

Accounts

40. Proper accounts will be maintained by the Sector Officers and will be provided to the ADC(G) for finalization of the district accounts.

41. Accounts of Government Grants under heading "4-Relief" will be maintained by the District Revenue Accountant under the supervision of EAC(R).

Assistance to Sector Officers

42. The services of officers/officials of other departments will also be utilized in case of emergency at the control centre and in the field. They are required to give full co-operation to the District Administration in meeting the situation created by flood. Heads of the Departments will issue instructions to their staff to render all possible assistance to the Sector Officers, in charge of their areas.

43. The EAC(R) in-charge of flood Control measures will act as Officer on Special Duty (Floods) and will be responsible for remaining in close touch with the flood situation in the district and collection information for compiling a flood report. He will keep ready with him the list of important organizations and the telephone numbers. He will also keep in touch with the concerned officers and maintain close liaison with Brigade Headquarters to utilize their services whenever required.

44. Every officer employed as flood relief duty will keep a copy of this flood order readily available for his use so that on receiving a high flood warning he may be able to act immediately. In the event of his proceeding on leave he will be responsible to hand over these papers to officers designated in his place and brief him properly. He will take over these papers on return to duty. In case of transfer of an officer, the copy of the flood order should be given to his successor and the latter be properly briefed.

ORGANISATION CHART - PUNJAB FLOOD WARNING DISSEMINATION & RELIEF OPERATIONS

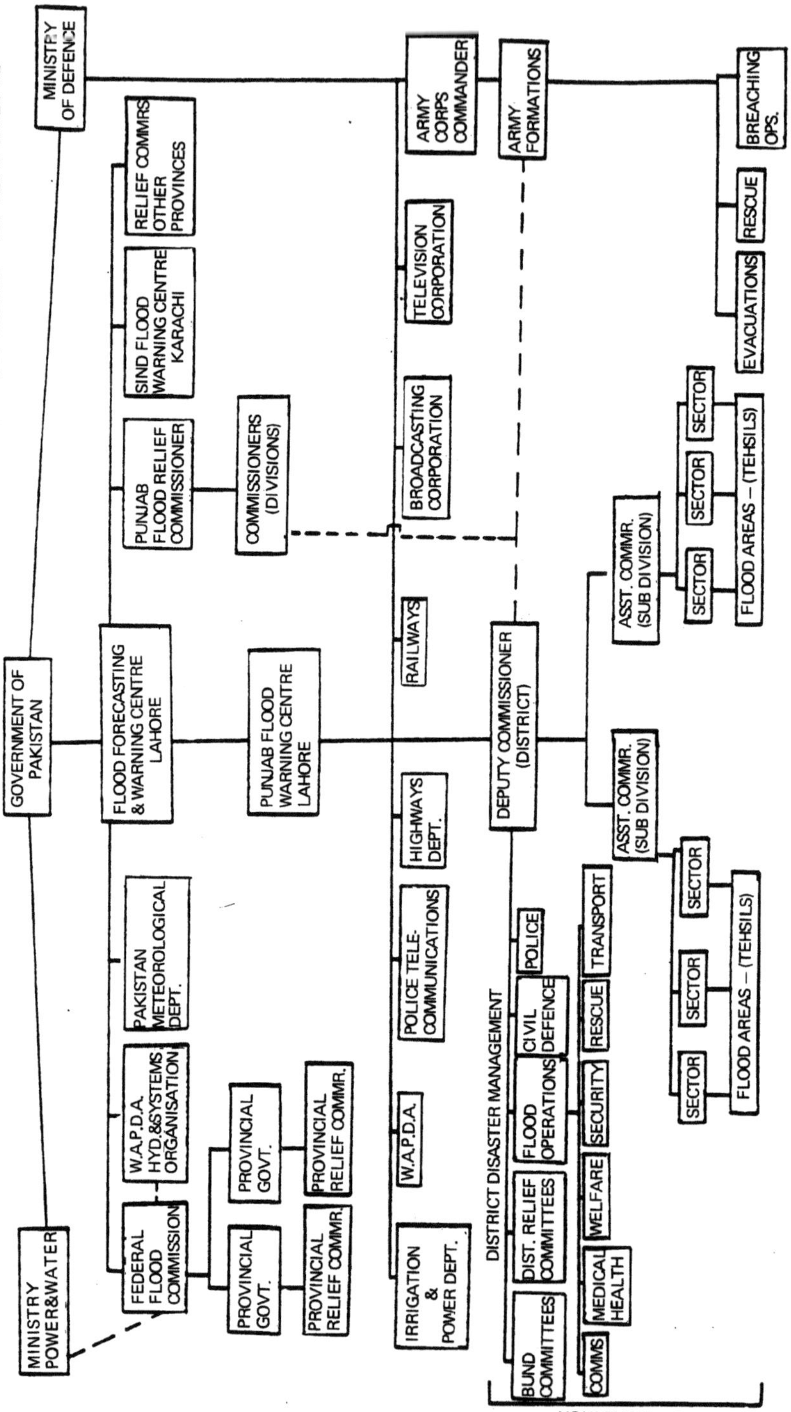

ANNEX 9

Weekly hydrological bulletin (Poland)

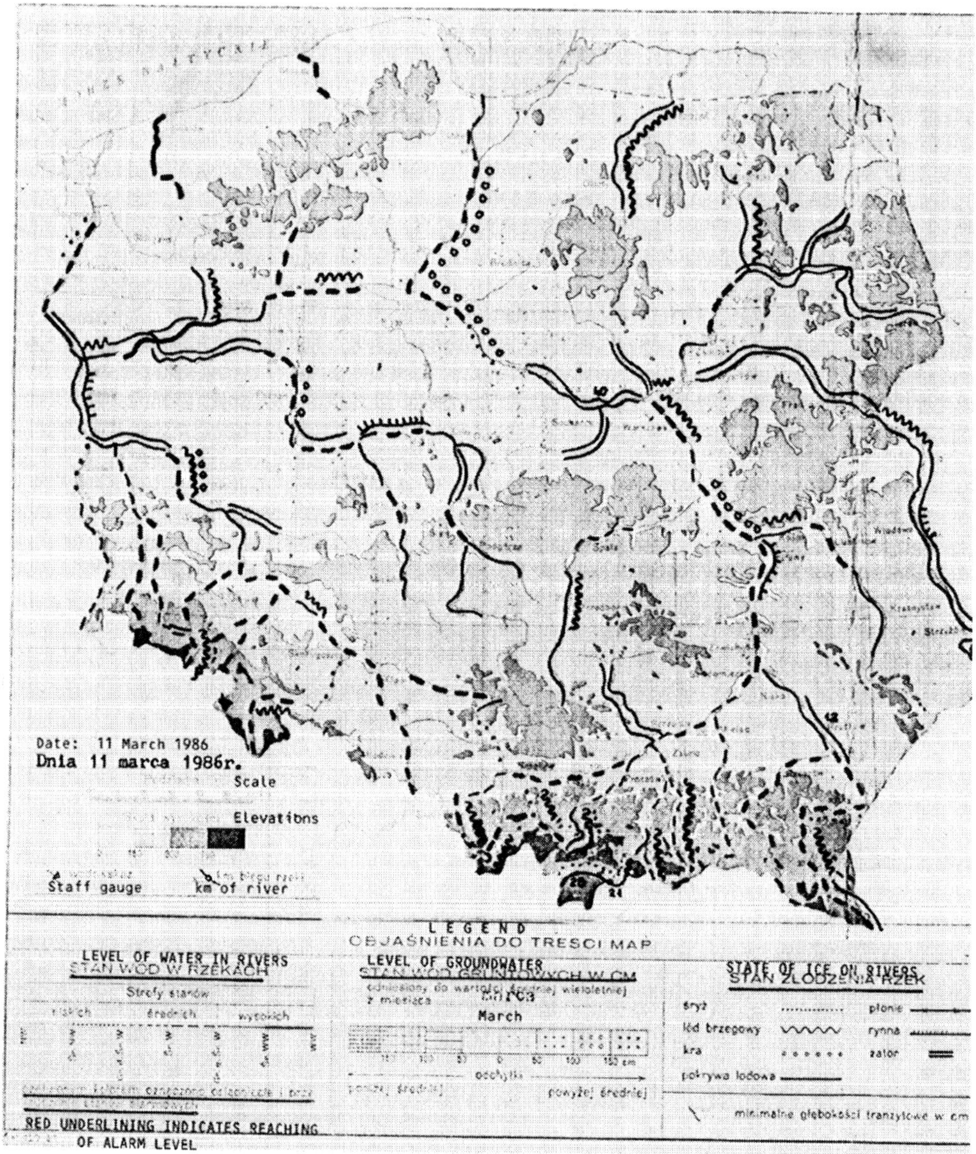
Date: 11 March 1986
Dnia 11 marca 1986r.
Scale
Elevations
Staff gauge
km of river
LEGEND
OBJAŚNIENIA DO TREŚCI MAP
LEVEL OF WATER IN RIVERS
STAN WÓD W RZEKACH
Strefy stanów
niskich
średnich
wysokich
LEVEL OF GROUNDWATER
STAN WÓD GRUNTOWYCH W CM
March
odchyłki
poniżej średniej
powyżej średniej
STATE OF ICE ON RIVERS
STAN ZLODZENIA RZEK
śryż
płonie
lód brzegowy
rynna
kra
zator
pokrywa lodowa
minimalne głębokości tranzytowe w cm
RED UNDERLINING INDICATES REACHING
OF ALARM LEVEL

SUM OF PRECIPITATION (MM) 5.3-11.3.86. DEPTH OF SNOW COVER ON 11.3.1986.

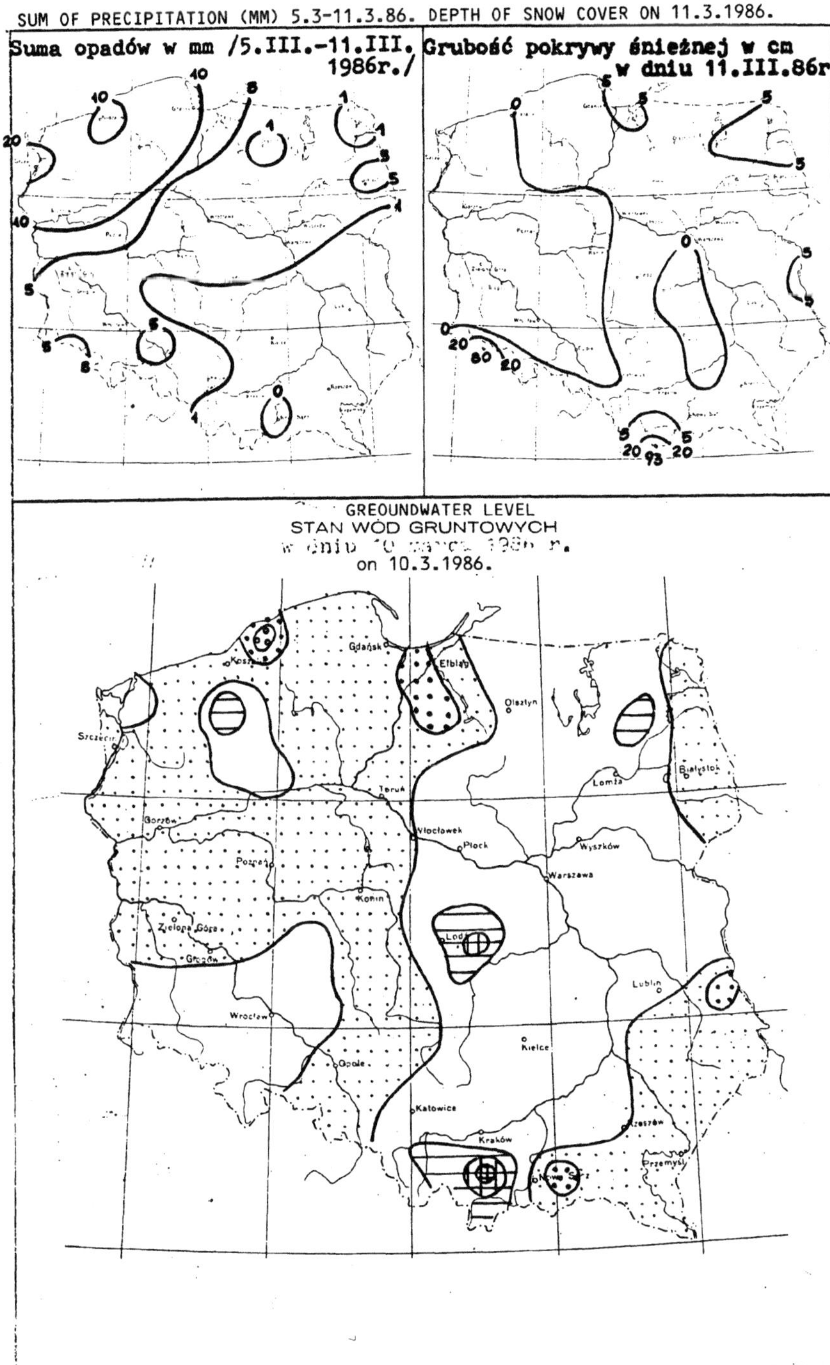

Projekt. inż. Jan Kostrzewa

CILSS / PNUD / OMM
PROGRAMME AGRHYMET

CILSS / PNUD / WMO
AGRHYMET PROGRAMME

CENTRE RÉGIONAL DE FORMATION ET D'APPLICATION EN AGROMÉTÉOROLOGIE

ET HYDROLOGIE OPÉRATIONNELLE

ANNEX 10

CENTRE AGRHYMET NIAMEY
Direction des Activités Opérationnelles
Division Analyse des données

River Niger at Niamey
Situation on 10 February 1986
Forecast of the decrease of flow

Situation

The rising part of the flood hydrograph of the Niger at Niamey is characterized in this hydrological year by high flows, superior to that of an average year (see Fig. 1) and approaching the median long-term values (see Fig. 2). The peak (1490 m^3/s) is however low. It is about the same as the peak of the dry year 1973/74 (1470 m^3/s) and is about eighth in an ascending rank from the lowest value of peaks, well below the median value of 1830 m^3/s.

Another important characteristic of the annual peak is its timing. It occurred on 15 December 1985 a rathr early date. Since the beginning of observations in Niamey in 1928, the peaks occurred twice in March, 22 times in February, 18 times in January and in the driest years since 1973, seven times in December and once in November (1984).

The limb of the hydrograph, which started this year rather early is thus characterized by a pronounced decrease of daily flows (Fig. 1), it resulted in January in an average monthly flow of 1974 m^3/s, near to the low flow of this month of ten-year occurrence. On 10 February the daily average flow was 414 m^3/s. For comparison, this of 1964 (average year) was of 1955 m^3/s; this of 1974 (year with the same peak) was 674 m^3/s and this of 1985 (absolute observed minimum) was of 189 m^3/s.

Forecast

On the basis of the different flow occurrence dates and their magnitude we endeavoured to forecast the evolution of the hydrograph (depletion curve) of this year using two mathematical relationships.

The first one permits to estimate the decrease of daily flows down to 100 m^3/s. It is based on the assumption that the depletion curve at Niamey is not influenced by precipitation and is exclusively depending on the rate at which the water in the river bed upstream is depleted. The exponential empirical relationships is:

$$Q_t = Q_o \exp(-W.t) \; [m^3/s] \text{ where}$$

Q_t = flow at time t

Q_o = the initial value of flow at the beginning of the time considered

W = depletion coefficient.

For this hydrological year Q_o = 1210 m^3/s (daily flow on 8 January), two values of W = 0,035 and 0,04. With these values it was forecasted that Q_t = 100 m^3/s should occur on 28 March 1986.

After this date the depletion was estimated by a polynomial of third degree

$$Q_t = Q_o + At + Bt^2 + Ct^3 \; (m^3/s).$$

Q_t, Q_o and t as above, A, B, C, = coefficients calculated from 11 years of observation from 1973/74 to 1983/84.

The last relationship estimates the flow of 10 m^3/s to occur 60 days after this of 100 m^3/s, e.g. between 11 and 27 May. The flow in the river bed could stop between 11 and 27 June 1986, if the rains over the basin would come late as they did in 1985, when a total stop of flow occurred.

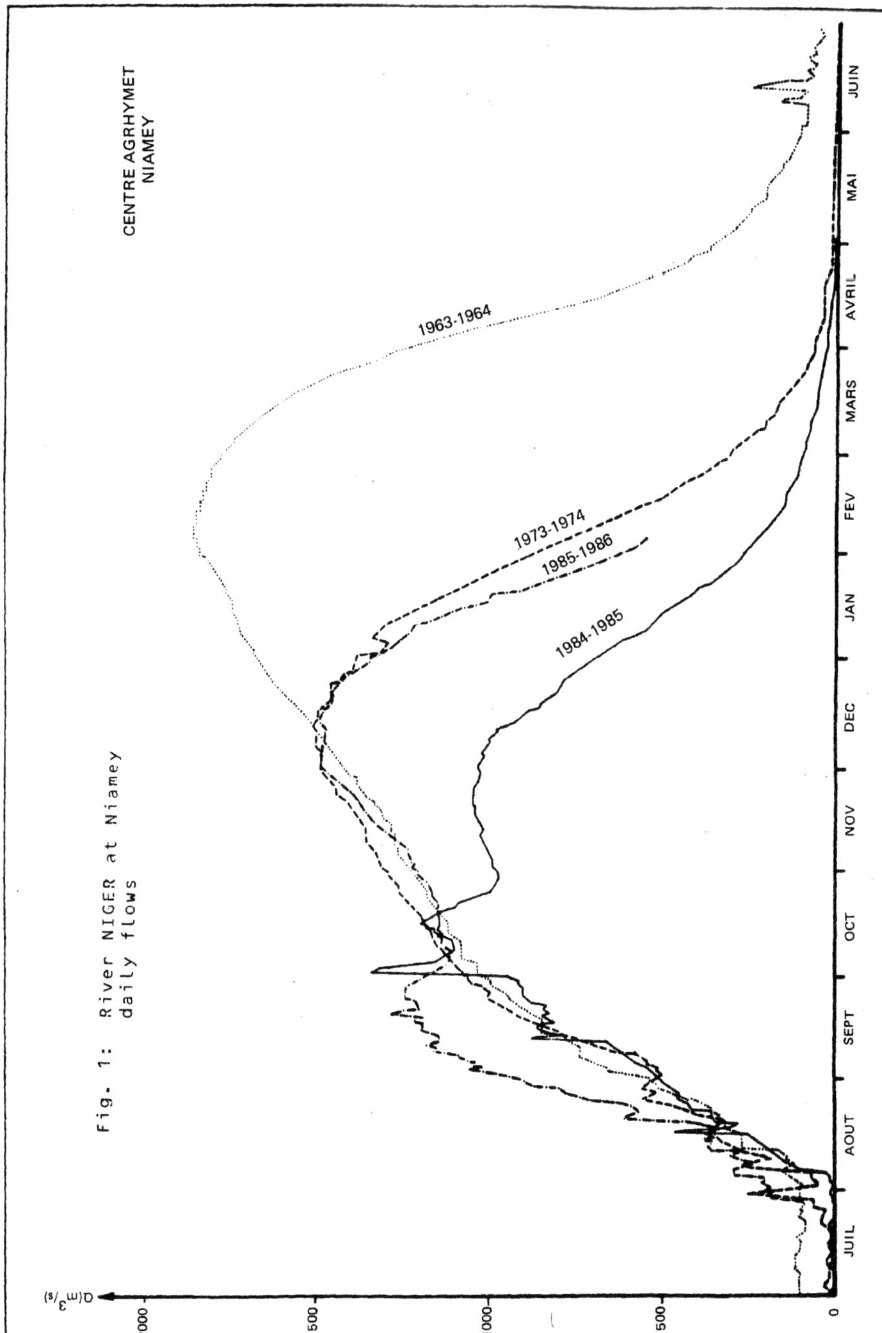

Fig. 1: River NIGER at Niamey daily flows

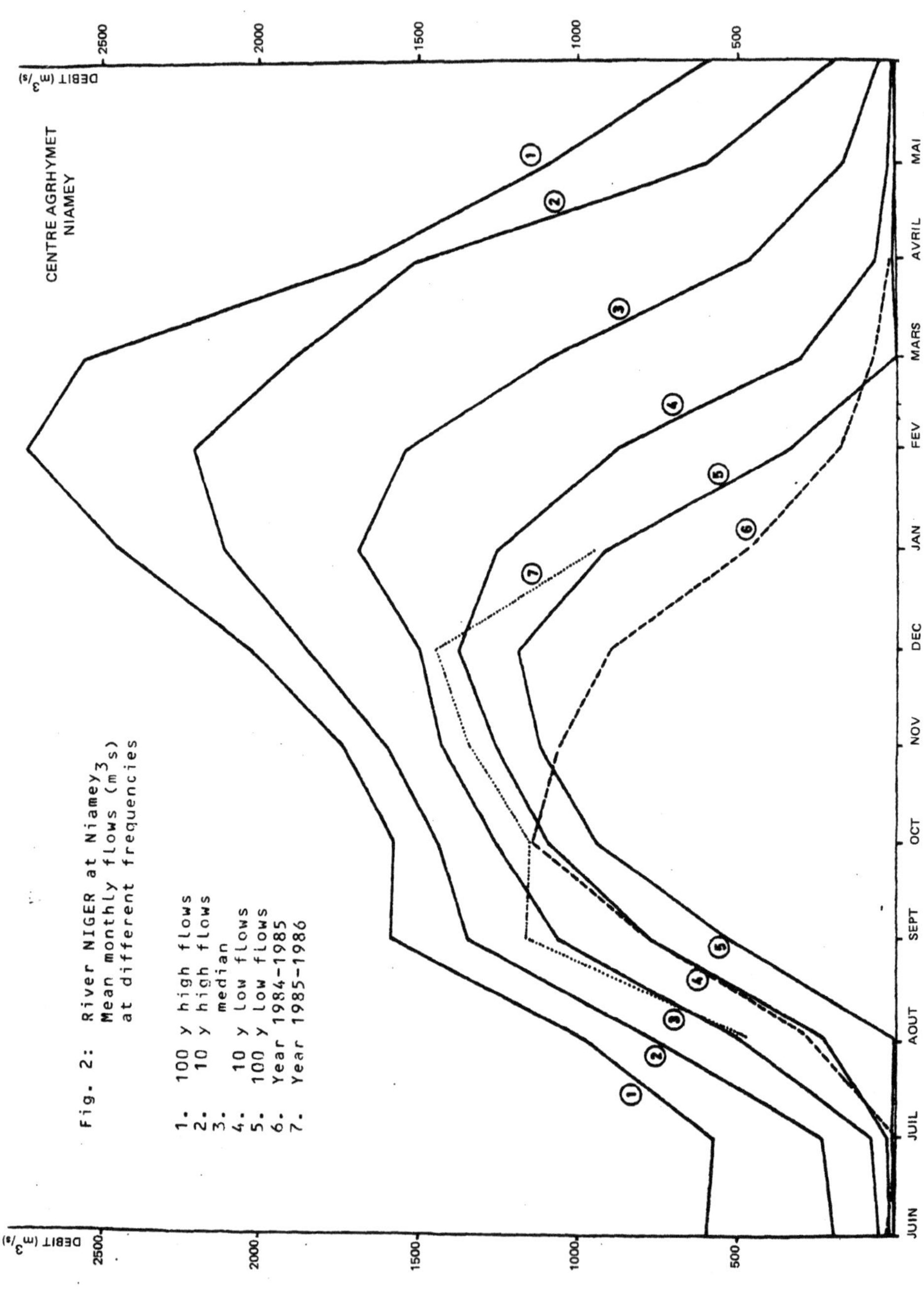

Fig. 2: River NIGER at Niamey
Mean monthly flows (m^3/s)
at different frequencies

Glossary of hydrological forecasting terms

Alarm level (stage)	Water level (stage) at, or approaching, flood level which is considered to be dangerous and warnings should be commenced.
Crest-stage relationship	Simple correlation(s), between corresponding flood peak stages at upstream and downstream stations.
Forecast (warning) lead time	Interval of time between the issuing of a forecast (warning) and the expected occurrence of the forecast element.
Forecast updating	Adjustment of forecasts of events as new information becomes available.
Hydrological forecast	A statement of expected hydro-logical conditions at a specified period and specified time.
Hydrological warning	Emergency information on an expected hydrological phenomena which is considered to be dangerous (GHP)
Ice phenomena forecasts	A statement of expected ice phenomena behaviour of a (hydrological) system due to various inputs.
Lateral inflow	Inflow of water to a river, lake or reserovir along any reach from the part of the catchment adjacent to the reach.
Long-term hydrological forecast	Forecast of the future value of an element of the régime of a water body for a period extending beyond ten days from the issue of the forecast. (GHP)

Medium-term (extended) hydrological forecast	Forecast of the future value of an element of the régime of a water body for a period ending between two and ten days from the issue of the forecast. (GHP)
On-line hydrological forecasting	The process of hydrological forecasting in which at least one of the components such as data collection, data transmission, data presentation and hydrological modelling are automated and/or computer-based.
Probability forecast	A quantified assessment (in real time) of future events giving the chance of its occurrence.
Response function	Mathematical representation of the behaviour of a (hydrological) system due to various inputs.
Seasonal hydrological forecast	Forecast of the future value of an element of the régime of a water body for a season (usually covering a period of several months or more). (GHP)
Short-term hydrological forecast	Forecast of the future value of an element of the régime of a water body for a period ending up to two days from the issue of the forecast. (GHP)

INDEX

GPSR Compliance

The European Union's (EU) General Product Safety Regulation (GPSR) is a set of rules that requires consumer products to be safe and our obligations to ensure this.

If you have any concerns about our products, you can contact us on ProductSafety@springernature.com

In case Publisher is established outside the EU, the EU authorized representative is:

Springer Nature Customer Service Center GmbH
Europaplatz 3
69115 Heidelberg, Germany

Batch number: 10607320

Printed by Printforce, the Netherlands